FLYING WITH FRED

FLYING WITH FRED

PENNY DE JONG

Edited by Edwina Harvey

Peggy Bright Books

Contents

Dedication — viii

Prologue — 1

One
My Father The Stranger — 9

Two
Why? — 17

Three
The Beginning — 21

Four
1940 — 25

Five
At Last! — 33

Six
Across the Pacific — 45

Seven
Canada — 51

Eight
Reality Approaches — 67

Nine
In the Desert — 83

Ten
From Sand to Fog and Rain — 97

Eleven
Hello Halifax! — 117

Twelve
The Realities of War — 129

Thirteen
Over the Mediterranean — 145

Fourteen
So What Happened to Fred? — 163

Epilogue — 175

Copyright © 2024 by Penny de Jong

All rights reserved. No part of this book may be reproduced in any manner whatsoever without written permission except in the case of brief quotations embodied in critical articles and reviews.

First Printing, 202

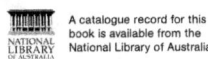
A catalogue record for this book is available from the National Library of Australia

Dedication

Dedicated to my father, Flight Lieutenant Frederick Bright, whose love of flying never dwindled.

To David Fowler, who opened my eyes to the wonders of aviation, with his own passion for aviation.

To my brother, Hugh Trevethan, also afflicted with the flying bug.

To my younger brother, Bill Bright, who reminded me of our childhood experiences,

To Heather, my Dad's baby sister, who was able to share family anecdotes with me.

To my family and friends who continually encouraged me to keep writing.

Prologue

Mum and I sat by Dad's hospital bed, just a few weeks before he passed away. At the age of twenty-seven, I was his only daughter, married with a baby girl who would never know her grandfather. Dad's skeletal frame jarred with my memory of the solidly built man of my childhood. With the loss of eight stone in weight, his appearance had dramatically changed. His now prominent cheekbones reminded me of the photos of him as a smiling young man in his blue R.A.A.F pilot's uniform. I regretted that we did not get the opportunity to grow the friendship which had evolved in recent years.

Mum's pale face and clenched jaw reflected her angst in anticipation of losing her husband. Her constant fiddling with her necklace irritated me, but I restrained myself from snapping at her. Attempting to mask her emotions, she valiantly chatted about anything but the bedridden dying man, who had been her husband of twenty-eight years. He would soon have his fifty-seventh birthday, and Mum would be a widow at fifty-five years of age.

Dad had always been an eloquent speaker, especially after a few beers. Yet he and I had always found communication between the two of us very difficult. I was shy of this man who told me very little about himself, and seemed to have a life at times entirely separate from the family. Yet, at the end of his life he was rendered mute although still very alert despite his serious illness. Strangely, I felt more connected to him on this day than I ever had in my life. Dad picked up his watch from the bedside table, and gave it to me, with a very intense expression.

"Do you want it fixed, Dad?" I asked, for once feeling as if I was of some help to him. Frowning impatiently, he shook his head, and gestured to Mum for pencil and paper. On the note, he wrote, 405073. The number meant nothing to me.

"What's that?"

Mum looked at the numbers. "That's his Service Number."

Then I understood. It was his R.A.A.F designation.

I looked at Dad and asked him, "Do you want me to get those numbers engraved on the back of the watch?"

Looking both relieved and frustrated, he nodded as emphatically as his weakness allowed. I too was relieved that I had guessed correctly, after feeling my usual squirmy sensation when trying to get things right with Dad.

I thought, "Oh well, if that makes him happy - it must be significant."

I had been born into a world of peace. To me, 'The War' was what I read about in history books.

The following day I returned the watch with 405073 engraved on the back. After examining it, Dad nodded with satisfaction and strapped the watch to his wrist. I did not understand how important this rite was to him then, but I do now, on writing my book. I was aware as a child that Dad had been proud of his service career, but I only understood his true passion for flying sixty years later.

Dad passed away two weeks later, one day after his fifty-seventh birthday. The funeral was family only with no flags, or military emblems to mark his contribution to a more peaceful world. My grandmother and aunts sobbed, while Mum, sitting next to me, shuddered, trying to suppress her emotions.

Several years later, Mum gave me Dad's old lawn bowls case, which contained his Flight logbook, letters from home, and service pay book. I was intrigued by war-time menus onboard a troop ship, and also one from a New York City restaurant. Foreign phone numbers were scribbled on scraps of paper. Were they from girlfriends? A calling card from London hostesses invited homesick servicemen for afternoon tea.

From time to time, I would sit alone, and carefully sift through these relics of his previous life, keen to learn more. I was grateful to have his logbook, but did not understand the entries. I knew that there was a wealth of information in this brown leather bowls case, but, for a long time, felt as if I was looking through a foggy window, not comprehending the significance. The child within me felt uneasy and tentative, sorting through what previously had been forbidden. Mum was very protective of these items, and only relinquished them when she was about to move to a retirement village.

2010

Over thirty years after Dad passed away, I met Dave, a close friend from my youth. While we were dating during the late 1960s, Dave would occasionally come to my parents' home for dinner. While I was helping Mum in the kitchen, Dad, as I learned much later, talked to Dave about his love of flying.

When I met Dave again in 2010, he told me that Dad's passion for flying had influenced him to train for his own private pilot's licence, and also to buy a 4-seat Cessna. I was amazed.

Over dinner one evening Dave said, "Mr. Bright showed me his logbook when I was at your house all those years ago."

Feeling slightly miffed, I replied, "Really? He never showed it to me! I do have it now, after Mum gave me his bowls case with the logbook and a lot of other stuff."

Dave smiled at me. "I would love to have another look at it now, after my own flying experiences."

I pulled the case out of a cupboard in the guest room and opened it, carefully unwrapping the logbook which was, by then, over sixty years old. With a sense of shy pride, I passed it to Dave. He looked carefully at the entries, respectfully turning the pages, his commentary slowly bringing the book to life for me.

"Those codes are the aircraft type and call signs. These others are standard aviation codes, such as doing circuits and landings, and emergency manoeuvres. See that red? That means he was flying in Operations. Look, that shows where the tail gunner was shot out."

Dave had my attention then!

"I wish I could know what it is like to fly a plane."

With a grin Dave said, "That can be arranged, you know."

"Really? I would love that - I think. I'd at least like to watch them fly." Although I had flown in commercial aircraft throughout my childhood, I had always wondered what it would be like to actually fly an aircraft.

The following weekend, Dave took me out to the local aero club, where his Cessna 180A was usually in a hangar large enough for three planes. There was also a make-shift kitchen and a couple of bunk beds for when he wanted to stay overnight. The back door of the hangar looked over a paddock with cows grazing, and a large lake in the background. Perfect setting for a sunset wine.

Dave's Cessna was being serviced, so not ready for flight. I was secretly relieved, knowing that I was 'safe' from the possibility of airsickness for the time being - or so I thought!

I was in for a shock as Dave quietly asked his friend George, who had a collection of vintage aircraft, to "take me up". At first, I was reluctant, remembering the last time I was in a small aircraft at the age of three. I went with Mum in an eight-seater aircraft on a flight between Brisbane and Bundaberg at low altitude. We had a lot of turbulence and I was extremely sick. My stomach still lurches at the thought of it. Consequently, when Dave had arranged my next small aircraft flight.I felt trepidation.

"But I've just had coffee!" I objected.

My protest was ignored, and I found myself in the front co-pilot's seat of a red, open dual cockpit vintage Fleet aircraft. Dave helped with my helmet and headphones. I felt obliged to continue, especially when Dave reminded me that the Fleet, manufactured in Canada, was one of the training aircraft that Dad flew in Quebec for his pilot training.

Dave's friend George said with a grin, "Do you know why the pilot sits at the back? Because the co-pilot gets killed first."

I thought, "Oh well - here goes."

Dave turned the screw, or propeller (prop). He yelled, "Clear!"

George, now in the rear cockpit, after preliminary take-off checks of the instruments, started ignition. As we trundled along the grass runway George said over the radio, "So far, so good!" Early planes such as the Fleet were 'tail-draggers', with either a ski or wheel under the aircraft tail. With the cockpit tilted higher, the pilot's view was partly obscured. George zig-zagged along the runway, using his peripheral vision to spot landmarks such as trees on the perimeter to keep his bearings before finally pushing in the choke, and pulling back on the control.

Within a minute we were airborne, flying above the lake, on a perfect

Queensland winter's day. I was smitten - the sensation was wonderful in such a small aircraft. To my surprise, I was totally unafraid - instead, highly exhilarated. I could feel the air rushing against my skin. The roar of the engine silenced any conversation, apart from over the radio.

Finally, a glimmer of understanding of how Dad had felt about flying, started to filter through to my being. That feeling has never left me.

Several weeks later, George persuaded me to fly with him in his Gypsy Moth. Jokes had been made about my taking the controls.

"No way!" was my response.

I climbed into the front co-pilot's seat of this rag-wing aircraft (the fuselage and wings are made of fabric, and then painted, facilitating a short take-off and landing) and waited for George to turn the prop, and climb in behind me. This aircraft had dual controls in both cockpits.

Before we started to taxi, George said, "My radio's not working. If I jiggle the joystick sideways, you have the controls. If I jiggle it backwards and forwards I'll take over."

"Oh, crap!" I thought, "My kids are going to kill me!"

Within a couple of minutes, we were airborne. Then, the joystick jiggled sideways. By then I was resigned to my fate: "If I die, I die - at least I tried."

Feeling surprisingly calm, I thought, "I guess I had better try this out."

Curious, I pulled the joystick gently back towards me - the Moth slowly climbed. Then, I pushed it gently forward, and it levelled again. (George later told me that he actually had kept control of the throttle, and rudders. I was so busy with the joystick, I hadn't noticed!)

Ecstasy flooded through me - this was better than waterskiing! I flew a circuit over the airfield and hangars, with Dave and others watching from below. After about ten minutes, the joystick jiggled back and forth, and I obediently put my hands on my lap, while George prepared for landing. I think I must have floated out of the Gypsy Moth that day, through sheer delight and wonder.

"Look at the smile on your face!" enthused Dave, taking one of his many photos of me. "Does this mean that you will like flying?"

I could feel my eyes sparkling as I breathed, "Oh yes!"

At last, I fully understood how much flying meant to Dad, and why he had been so restless all of his adult life He was permanently grounded.

Dad, this book is for you - your flying continues.

One

My Father The Stranger

Dad was an enigma to me until three years before his death. He did not act like the father I needed or wanted - his passions were outside home, with his work, the pub, or sport. He was a good-looking man, although not tall, at 5 foot 8 inches. Air force photos depict a slim, dark-haired man with a strong jaw, and a mischievous glint in his eye. A cigarette was usually neatly placed between two fingers, as a fashion statement. The Dad I knew years later was heavier, with jowls appearing and disappearing over the years, as he attempted to control his weight.

On workdays, he always wore a starched, bleached white shirt, laboriously ironed by Mum. For the era, he was a 'sharp' dresser, hair slicked back by Brylcreem. Shoes were always well-polished, also by Mum.

Yes, Dad was the king of his small castle. Weekends usually saw him driving the company car to Lawn Bowls in his regulation creams.

On ANZAC Day, he would rise for the Dawn Service, and put on a suit adorned with his medals.

On some ANZAC Days if by chance he decided to attend the mid-morning parade, Mum, my younger brother Bill Jr and I would walk into town, to watch the parade. Dad would march proudly past our viewpoint, chin up, and hands tucked to his side. He never acknowledged our presence, such was his military pride. He would then eventually roll home after many beers and reminiscences at the Services Club.

On those rare days when he was at home, Dad would slouch on the couch with no shirt, a pair of casual shorts, and rubber thongs. Family times seemed to be a chore to him. On the occasions when we did manage to have family outings such as going to the 'pictures' or Sunday afternoon drives, when we would often go down the range from Toowoomba to the small town of Grantham, where Dad would go into the pub. In those days, women and children were not allowed in the bar.

Consequently Mum, Bill and I would wait in the car. Dad would walk back out with a tray carrying a beer for Mum, and cherry cheers for Bill and me. He would then disappear for an hour or so, enjoying his beers at the bar. I would read my book, and Bill would whinge about being bored. Dad always let us know how generous he had been with his time.

He intimidated me from an early age, not giving physical affection, apart from perfunctorily presenting the side of his face for a peck on the cheek when I said good night. He never hugged me. Dad's idea of showing affection was that of physical teasing. After stirring his black coffee, he would place the hot spoon on my hand. Another favourite was a 'horse-bite' when he would flick his fingers on my thigh. Both gestures always resulted in my tears.

Dad would say, "Don't be such a sook! It was only horseplay!"

In the background, Mum would admonish, "Fred! Leave her alone! She doesn't like that!"

There was also verbal teasing or criticism - he would tell me that I was either too skinny, too slow, or too clumsy. My self-esteem suffered as a result. I believed that he did not love me.

Mum would say, "Oh Penny - of course he loves you! He is very proud of you and talks about you to his mates."

"Then why doesn't he tell *me!*"

I often felt 'squirmy' around him, not understanding why. If he walked into the lounge, determined to watch his favourite

TV show, my stomach would knot, and I would silently go to my room.

Dad was rarely home for dinner at the right time. Mum, trying to orchestrate a family meal, would phone the local pub, asking, "Is Fred Bright in the bar?"

Usually Dad would appear half an hour later, looking a bit sheepish, while often carrying the prize of a meat tray from the nightly raffle. That usually appeased Mum to a point as it helped with her food budget. After lawn bowls on Saturdays, a 'chook' was usually presented for Sunday lunch, the one day of the week when we did eat as a family.

Yet, I realise now, that there were times when he shared his interests, in his own brash way, with the family. He loved the music of Gershwin, Ella Fitzgerald, Glenn Miller, and all hits of that era. I asked about his early endeavours as an amateur photographer. I enjoyed looking at the photos he had taken and developed in a home-made dark room.

The only tangible evidence of his passion for planes, for me as a child, was my watching him at the kitchen table making balsa and plastic models of the Halifax, Mitchell and Lancaster aircraft that he flew during his service. I can still smell the woodcraft glue. Bill was included in this hobby because he was a boy!

On the night of my first ball just before my sixteenth birthday, Dad made a point to be home as I waited for my partner to collect me. I was almost shocked at his reaction when he saw me in a pink silk gown, with my hair in a French knot. He could barely speak, and his eyes glistened. I suddenly felt shy, moved by his reaction. I had been expecting sarcasm, as usual.

In 1968, at the age of nineteen, I started my nursing training, and was required to live at the Nurses' Quarters. Officially, I had left home. It was only then that I had an inkling that Dad was feeling a mixture of remorse and guilt at his emotional absence during my childhood. He started calling me his "number one daughter". Although I did not admit it to him or Mum, I was secretly touched by this small sign.

Mum and Dad eventually moved to Brisbane from Toowoomba for work. I continued my training in Toowoomba, catching the bus to Brisbane every couple of weeks on my days off duty.

Dad would meet me at the bus station to take me home. While waiting for me to arrive, he would have a few beers at the local pub. One evening, he arrived at the bus station, obviously drunk and emotional. I had that squirmy feeling again. Had Dad been thinking of his lost opportunities to have a real connection with me? I was a captive audience in the car, while he told me how proud he was of his 'number one daughter', and for the first time ever, announced that he loved me. I was twenty years old. At that moment, I decided that his declaration did not count, as he was drunk. I still did not believe him, despite Mum trying to convince me later. My child-self still grieved for the absent father.

A year later, while training for midwifery in Brisbane, I met John, who would become my husband. He could give me all that I had craved throughout my life until this point: physical affection and emotional security. All went well when he first met Mum and Dad. But that changed very quickly when Dad discovered that John was Dutch, and Catholic. At the time he also had a beard.

I had no idea that Dad was so biased against Dutch people, or Catholics. Mum explained that Dad's family had an issue about an interfaith marriage between Catholics and Methodists, causing long-lasting bitterness within the Bright family. Until then, I had been totally unaware, as the issue had never been discussed. In addition, Dad had served along-side Dutch aircrew during the War and found them arrogant.

After I had been dating John for six months, he brought me home from the movies one evening. Dad met us on the back porch. He looked at John and said, "I'm telling you that I hate the Dutch, I hate Catholics, and I hate beards!"

I was furious. For once finding my voice, I shouted, "You can't say that to John! We're getting married, and I don't need your permission! I am over twenty-one!"

From that moment, everything started to change. I had

discovered my own strength, and Dad was starting to feel a sense of powerlessness

Mum was torn between placating me while staying loyal to Dad.

When John and I announced our engagement, Dad refused to give us his blessing. Up until two weeks before our wedding in 1973, he had refused to give me away. When he realised that I was not going to give in to his objections, he succumbed, his pride broken. As I was the only daughter, this was his only chance. Sadly, he did not live to see Bill married.

I can still hear the wedding photographer imploring, "Smile, Mr Bright!". Dad did not smile. At some level, I understood how very difficult this was for him, but I was happy that both my parents saw me get married. My fears of a drunken, ranting speech at the reception were unfounded - Dad's speech was sober, eloquent, and well-meant.

John and I escaped to New Zealand for a year to take some time for ourselves. On our return, Dad, who had slowed down his drinking, extended his hand to John and apologised for his behaviour, welcoming him into the family.

"You will make a good husband for my daughter."

I nearly passed out, unsure how I should feel at such a turnaround in Dad's attitude.

From then onwards, Dad went to a lot of trouble winning John over by showing him his model planes and woodwork projects. A sense of friendship and mutual respect was born for John, and also for me. My anger and bitterness towards my father lessened, but I had not forgotten.

In 1976, when our first daughter, Carol, was born, Dad became ill with cancer. He held his granddaughter just once, calling her 'threepence'. Mum's nickname was 'tuppence' so the name fits. From that time until Dad passed away five months later, my affection and respect for him grew as I watched him struggle valiantly with an ugly illness that took away his speech.

Mum had great difficulty dealing with this disability, but, for me, the silence was a stronger form of communication between Dad and me than when he was well. Somehow, we finally understood each other, without saying anything.

After Dad passed away, he came to me in a dream, telling me, "I'm alright." For weeks I would look for him in the street, sometimes thinking that I saw him.

Although I had reached a point of acceptance of my childhood circumstances, I would need more life experiences to equip me with a greater insight into my father's ways.

Two

Why?

Why was Dad so restless and unsettled? It was more than his stubborn and defensive personality. His father was often depressed, and his mother fussed over her only son, taking care of his meals and washing until he joined the R.A.A.F in 1942, at the age of 22.

Dad always had a rebellious streak and did not like conforming. It was most likely his parents' discipline that kept him out of serious trouble.

From what I know, he hated school, apart from Maths. During my childhood, Dad did tell me that he had been bullied at the age of 14, at school. When he was swimming in the pool, a bigger boy tried to strangle him with a chain. Dad retaliated by nearly drowning him.

I don't know of his demeanour before he enlisted in the air force. He was very determined to join up, despite his father's objections. Following his 21st birthday, Dad succeeded in having his father sign the consent form.

Photos and postcards sent home during the war years indicated that Dad was enjoying life in the air force. He could follow his passion for flying and enjoy some personal freedom on leave.

Dad joined R.A.A.F 462 Squadron Middle East Command in 1943 after two years of training. As the squadron was attached to the R.A.F the Australians were forced to prove their worth to the British contingent, with its disdainful attitude toward the "colonials", who were regarded as second-class citizens. Dad flew his 30 missions out of the Middle East, and then a few more from Malta.

Out of the conflict zone, he trained new pilots, and then flew transport aircraft within Australia, before demobbing in 1946. Did this give him a chance to settle any nerves, before returning to his old job in civilian life?

Dad and Mum met and married in 1948. I was born the following year. Apparently, he was restless from the start of their marriage, with Dad just going to the pub after work, and leaving Mum alone with a baby and domestic chores. He would eventually return home late each night to a disgruntled wife. His interest from those early days was more drinking with his mates, some of whom had also been in the armed forces. Did he talk with them about his wartime experiences? He never talked about it with his family, who did not understand the trauma of war.

> *"There is a world of silence between those who have seen unspeakable acts, and those who have not." (Professor Sandy Macfarlane).*

The family felt left out, disregarded, trying to understand the impact of being on the front line had on him.

Dad's younger sister told me that she remembered a time after Dad had been demobbed and returned to his family home that he was sitting in the bath. A knock on the front door announced the arrival of one of Dad's tail-gunners from his crew. Dad called out for him to come into the bathroom, where he sat on the edge of the bath.

The family clustered outside the door, listening to the only conversation they would ever hear from him about life in a Halifax over Europe.

Otherwise, Dad would be very withdrawn and non-communicative about his war experiences, apart from answering my occasional childhood questions when he was in a good mood.

I have tried to ascertain if he had post-traumatic stress, and suspect strongly that he did, although, as far as I am aware, he showed no signs of reliving conflict episodes, by jumping at loud noises, or, as far as I know, having nightmares.

I suspect he coped with his experiences by blocking them out with alcohol, of which he was a little too fond. He never acknowledged that 'the war' had affected him, but Mum explained to me many times that it did. I was a child, and all that I knew was that he was not behaving in the way other fathers did. I blamed him, and I blamed myself. Now, of course, I know it had nothing to do with me, but everything to do with the fact that post-traumatic stress was not recognised until after the two World Wars. There was no support.

After four years away at war, Dad, like so many others, was expected to go back to his civilian life as if he had never been in combat. Consequently, he outwardly complied, while battling with his inner demons.

How did he get to this point? Before I try to work out the details of combat stress on Dad, I will go back to the beginning of his life.

Three

The Beginning

Fred was born into the new era of aviation. His childhood would have been one of excitement, as he saw the new Qantas Havilands fly over his house, en route to Brisbane.

"Mother, may I go to the pictures on Saturday? Please?"

"Only if you have done your chores and your homework."

Saturday was the highlight of Fred's week. If he could go to the pictures, he knew that he would see the news about his heroes, Charles Kingsford-Smith, Bert Hinkler, and Amy Johnson. He just *had* to do what they were doing!

On the days he did not have enough pocket money for the pictures, Fred would ride his bike to Wilsonton Airport, about thirty minutes away - twenty if he pedalled fast! Chaining his bike to the boundary fence, Fred would find a spot in the long grass, so that he could see any planes that may have been taking off or landing. Sometimes, he didn't see any, and he would trundle home slowly in disappointment. If he saw a Gypsy Moth, he would lie on his stomach in the grass, enthralled. The burble of the engine, and the smell of the fuel overwhelmed him with passion.

For his 12th birthday in 1931, his maternal grandparents Anna and George, who doted on their first grandchild, gave Fred *The Wonderbook of Aircraft*. The thick, hard-cover book contained photos, drawings, and text describing the early history of aviation. That book has survived over seventy years. Fred obviously cherished it, as it is still in exceptionally good condition, apart from a crayon drawing by a young artist - most likely me - many years later.

In 1932, Fred's Nanny and Grandad took him out to the airport, to watch Charles Kingsford Smith land his Southern Cross, a Fokker Trimotor. Even at such a young age, Fred already understood the fundamentals of aviation. His skin tingleed at the thought of being up there, at the controls. Smithy was Fred's hero. The beloved *Wonderbook* was tucked under his arm.

Surging forward, Fred managed to find a place exactly where Smithy was about to jump down from the cockpit. Smithy noticed the young boy with an eager face, pushing his book towards him. Several other boys had the same idea. In a perfect copperplate hand, Smithy wrote;

C. Kingsford Smith
Southern Cross

A later hand had written; *7.8.1932.*

Thirty years later, I remember Dad taking me to Brisbane airport, where the Southern Cross lay in its display hangar. I must confess that, at the time, I did not really appreciate the reason behind the gesture. He was trying to share his passion for aviation with me, but my protective walls had already been built.

Fred's family had already felt the effects of World War I. Clara, his mother, saw her brother and two uncles leave for Gallipoli, Egypt, and France. Miraculously, they all came home. However, they were 'different' for the rest of their lives.

Fred's father, Bill, at the age of eighteen years, had enlisted in the army. He never served overseas, instead working as a telegraphist in Brisbane. His service records describe a 'nervous condition' which resulted in a medical discharge a few months later. Bill then got work as a telegraphist for the railway, married Clara in Toowoomba, and Fred was born soon afterwards.

Only twenty years after World War I ended, those who had endured the austere war years now had to face a second global war. The viewpoint of those who had been there before with conflict, would have been the total opposite of the next batch of gung-ho, enthusiastic young men looking for adventure. In the worst way, history was repeating itself.

During his youth, when Fred was not living and breathing the new world of aviation, he attended a state school, and then Toowoomba Grammar, thanks to Nanny. He left at Junior Level.

While working as a clerk for a motor firm, he studied bookkeeping at the Tech. In his spare time, he would follow his interest of photography.

When World War II was declared, Fred, then aged twenty, knew that he would have to wait before he could join the air force, due to Bill's strong opposition. By the time the opportunity to enlist arose, there would be no stopping Fred. He had waited long enough.

Four

1940

Hume Street, Toowoomba

"Is that you, Snowy?" (Fred, my father, had *Snowy* as a nickname).

"Course it is. Mother, I have something to tell you all. Where are the girls?"

"Kathleen is out, and Heather is doing her lessons. Your father is in the backyard."

"Heather, come out here - I have something to tell you all."

Clara, feeling a knot in her stomach, took the saucepan of mutton stew away from the heat of the wood stove, and, after calling her husband Bill in from the back yard joined her family at the kitchen table.

"We might need a beer tonight, Dad. Mother, you could have a sherry."

Thirteen-year-old Heather, slippers flopping up and down on the lino, walked down the hall, into the kitchen.

"What can I drink"?

"Cordial!"

Bill's eyebrows arched. "Well?"

"It's happened – I've been called up for the air force. I start training in two days. Remember last week when I got a lift with Ken to Brisbane? It wasn't to see a girl – I was signing up."

Apart from Clara's sharp intake of breath, a cold silence descended on the room.

Bill glared at his son. "Haven't we had enough of war in this family?!"

"Dad, you wouldn't give consent before I came of age, but I'm going on twenty-two now, and I can do what I want." Fred's jaw jutted out in stubborn determination. He looked at his father, waiting for a reaction.

Bill said nothing, picked up his tobacco pouch and cigarette papers, and walked out the back door to sit on the wood heap. His dark mood had descended.

Heather's mouth opened and closed again. She was both stunned and proud of having someone in the family fighting for the cause. It was all the talk at school.

Clara was weeping. "Oh Snowy! I don't want you to go! You might not come back! What about your job with Redman's?"

"They promised me that my job will be there when I get back. Mother, I will come back - the war will be over by Christmas. I'll be training in Brisbane for a few months, so I can come home on leave."

Clara's weeping got louder as the news penetrated. Heather joined her in sympathy.

Fred pushed his chair back forcefully and stood, hands leaning on the table.

"If you're all going to be so bloody miserable, I'm going down the pub for a quick beer before the six o'clock swill."

His heavy footsteps echoed down the hallway. The door slammed as he left

The stew went cold on the stove. Nobody ate dinner that night.

It was now evident that Fred's burning desire to fly at any cost was so strong.

Clara insisted on helping Fred to pack his belongings. She had made a boiled fruitcake and cut some sandwiches for the train trip to Brisbane. The night before Fred's departure for Brisbane, Clara cooked a special roast dinner, with sago for dessert.

Fred dressed himself in his Air Cadet's uniform for the occasion of his departure . Suddenly, he looked taller, but to his mother's eyes, still so young!

For once feeling in charge at home, Fred organised to have a family photo taken after dinner. He set his camera for a one-minute delay and posed everyone before jumping into the scene before the shutter clicked.

The photo of a family of five tells its own story. Fred and his sister Kathleen stand behind Clara. Heather and Bill, who are seated. Only the youngest, Heather, has a faint, forced smile. Older daughter Kathleen is serious, dressed in dark clothes. Clara, hands clenched in her lap, looks as if she has been weeping. Bill is glaring at the lens. Fred's expression shows a combination of pride, sadness and determination. Perhaps reality was starting to dawn on him.

On the morning of Fred's departure, the whole family piled into Bill's Model T Ford. Still in charge, Fred suggested, "Let me drive, Dad - it'll be quicker."

Reluctantly, Bill agreed. He looked defeated at his inability to keep his son out of the war.

Clara, clutching her handkerchief, was trying hard not to cry again.

"Heather! Will you stop wriggling!"

Too soon for everyone but Fred, the Ford pulled up outside the railway station.

Fred grabbed his bags and kissed his mother and Heather goodbye. As Kathleen had to leave for work earlier, she and Fred had said their goodbyes at home.

"Heather, stay in the car!" snapped Bill.

"Humph!"

Heather crossed her arms across her chest and sulked.

Fred walked with his parents onto the crowded platform. Other families were also saying goodbye to young men ready for adventure. A mixture of excited chatter, laughter, and some sobbing could be heard.

Why did Fred have a tight feeling in his throat? Thankfully, he saw some friends, also enlisting, boarding the same train. With a brief glance at his father, Fred extended his right hand. "Cheerio, Dad!"

Bill cleared his throat, and merely nodded as his son climbed aboard the train.

Clara was crying and waving her hanky. Bill cleared his throat again and nodded briefly at a porter he knew.

With a belch of soot and steam, the train tooted, and started chugging slowly out of the station. Fred leaned out the carriage window, with a final wave.

Slowly and sadly, Bill and Clara traipsed back to the Ford. Bill patted his pocket, "Where are the keys?"

Heather piped up helpfully, "Fred has them!."

"What! Why didn't you tell me!"

"You told me to stay in the car!"

Bill sent a telegram to Helidon, the next station, alerting Fred to hand in the car keys for the train's return trip to Toowoomba.

Fred had just begun what he thought would be a life of flying and travel.

Clara and Bill had just begun four years of waiting and hoping that the pink telegram would never arrive.

Front row: Clara, Heather and Bill Bright. Back row: Fred and middle sibling Kathleen

Five

At Last!

In 1939, the British Commonwealth took steps to provide allied air forces in order to support the British R.A.F. Australia joined the Empire Air Training Scheme, which provided basic training in Australia, before cadet pilots undertook more intense training in either Canada or Rhodesia before moving to the British Isles.

March 1941

Six long, sooty hours after leaving Toowoomba, Fred arrived at Roma St Station Brisbane, where he and the other recruits changed trains and travelled a further hour to bayside Sandgate, the base of the No 3 Initial Training School. He would have to wait for fourteen weeks before he got even close to an aircraft. First, there were military drills, and settling into a vastly different life, sharing space in barracks with fellow recruits. When not drilling or exercising, the recruits studied mathematics, navigation, and aerodynamics. Each night before falling asleep, Fred would rehearse his growing knowledge of aviation, willing himself to be 'up there'.

> *Dear Mother, Dad, and Girls,*
>
> *I am in camp at Sandgate, with the other new lads. They look like a good bunch. There are forty men who want to fly, but only thirty will be accepted in this round. As you would know, I intend being in that thirty. At first, I felt as if I was back in Scout Camp, but they haven't wasted time shaping us up. We have all been poked, prodded, and jabbed for dysentery, syphilis, and typhus. I am now officially healthy. The barracks are basic corrugated iron huts with rows of camp beds, and palliasses stuffed with straw. For the rest of the first week, we had the chance to break the ice with one another and get orientated with the place, and a new routine. The second week was a shock!*
>
> *We are woken by a very loud sergeant at sparrow-fart, at half past five every morning. Beds must be stripped, and palliasses rolled up, ready for inspection.*
>
> *Rain or shine, we are on the parade ground at six thirty for a full parade when Orders of the Day are announced by the Warrant Officer on Duty.*
>
> *Breakfast is in the Mess at 7 o'clock. If you are late, you will miss out! The porridge is not as good as yours, Mother!*
>
> *After breakfast, we start the day's classes. Lucky for me, I*

find the maths easy, and have a basic understanding of navigation - could come in handy later!

Every afternoon, we do bloody parade drill! Up and down, up and down. Buggered if I know what marching in time, arms raised to shoulders in the right order has to do with flying!

I must obey orders and do my best if I want to fly. We get leave in five weeks, after the initial training. I will let you know the date when I get a pass, and a time for the train. Looking forward to seeing you all again.

Your son
Fred

P.S. Mother, when not on parade, we wear khaki overalls that fit like hessian sacks. The navy dress uniform is for going out!

June 1941

Fred got his six-day leave pass at the end of his initial training, and returned home to Toowoomba, much to the relief of Clara.

"Son! You've lost weight! Your hair is so short!"

"The mess food is adequate, Mother, but not as good as your cooking!"

Clara smirked proudly.

"What have you been doing, Fred?" Heather was curious to learn more about this new adventure in the family.

"Well, it's a bit like boarding school at Grammar, except we get paid. Rules, rules, and more rules. Look out if you don't obey! I like the exercise drills, but marching is boring. I want to fly, not march! I knew a lot of stuff about aviation before I started, but the course became a lot harder in the last two weeks. Three air cadets failed the exams, so they are being posted to signals. They're pretty cut up about that. Glad it wasn't me."

Bill walked into the kitchen from the backyard, so the conversation stopped. Heather muttered something about homework and went to her room. Fred looked at his father's stern face.

"So, they haven't thrown you out yet?"

"No - that will not happen, Dad. I am serious about this."

"Humpph!" Bill picked up the newspaper and left for the outhouse.

Fred looked at Clara, whose wavering lip warned him that there would be tears very soon. He kissed her on the cheek

"I'm going out for a while, Mother - I want to call in and see Kev Fallon. He's really keen to enlist, but his old man won't sign the papers. So he has to wait another year probably."

"Poor Mrs. Fallon," murmured Clara as the front door slammed.

At the end of his leave, Bill drove Fred to the station. This time, the farewell was much calmer as Clara knew that her son would be safe, at least for a few months. He would be home again in a few weeks.

On 1st June 1941, Leading Aircraftman Fred Bright commenced his Elementary Training at west Brisbane's Archerfield Aerodrome, where two training squads were formed. At the time, Archerfield was also Brisbane's prime civilian airport. Consequently, trainer pilots were sharing airspace with more experience civilian pilots. This would make life exciting for some, and annoying for others.

The training aircraft were DH 82s or Tiger Moths, with dual open cockpits. Designed by Geoffrey de Havilland, the "Moth", as it is affectionately called, needed a 'prop start' by groundcrew, with the pilot at the controls from the rear cockpit. In aviation jargon, the Tiger Moth was a 'tail dragger', having a tail wheel, instead of a nose wheel. Taking off and landing are tricky, as the nose rises when on the ground, impairing pilot visibility. This is managed by taxiing in a tacking movement, side to side, to get a clearer view. Landing must be on all three wheels at once, so that is a skill within itself.

June 2nd saw Fred in the front cockpit of his first Tiger Moth-A173. Flying Officer Reginald Robinson would be one of the most important people in Fred's life, as he trained this cadet, and many others. Reg was easy-going, but strict on technique and safety.

Fred found the opportunity to ask his instructor about his own aviation experience.

"I was seconded from civilian flying. Have been flying Puss Moths (a three seater- high wing aircraft similar to the Tiger). My father is the founder of New England Airways. So, that was an easy way to get my license. I'm too old at thirty-two to serve overseas at this stage, unless all you young blokes get killed off."

Fred felt a slight shudder at that comment.

His first time 'up' in a Tiger Moth was all that he had anticipated. Strictly as a passenger this time, Fred could enjoy the sensation of taxiing, climbing and finally soaring over the local Archerfield area. He was breathing heavily with excitement, pulse racing. He could feel the wide grin on his face. The horizon dipped as Reg prepared to do a circuit in preparation for landing.

"Better than sex!" thought Fred.

During the thirty-minute flight, he kept an eye on the

rudders and 'stick' in his cockpit as they mimicked the actions of the pilot.

On the following day Fred did not leave the ground, instead learning the controls, and how to use them for take-off, flying, and landing. This was a skill that would need to be learned and practised again and again. His life and future career depended on it. For the next two weeks, Fred flew with Reg Robinson, learning and practising circuits and landings, twice daily.

On his seventeenth Circuits and Landing, Reg yelled to Fred through the Glasson speaking tube, 'When we land, I am getting out. You can take her up - just one circuit and come back down again. Just do what we have been practising. Enjoy! Don't wreck the aircraft!"

With a mixture of excitement and anticipation, Fred, now on his own, completed a text book take-off. At last! This was great!

"Shame I can only do a short circuit."

Knowing that there would be big trouble if he did not comply, Fred prepared for landing. A cross-wind buffeted the aircraft just as the wheels touched the ground. The Moth bounced down the runway, eventually coming to a stop.

"Good on ya, Bright!" the other cadets chortled.

"Don't worry! Tomorrow it will be your turn!' Fred retorted, as he walked away from the aircraft. (The best aviation incident is the one you can walk away from).

Fred would have been surprised and pleased to know that his daughter, fifty years later, had the same thrill of flying a similar aircraft, the Gypsy Moth - unforgettable!

On the last day of this training phase, he wrote in his log book:

Certified that I have been instructed and understand the petrol ignition and oil system of the Tiger Moth Aircraft and that I know that the safe endurance of this type of aircraft to be two hours and thirty minutes. 14th June.

These two weeks were crucial to the progress of the cadets - they would be declared air-worthy, or 'scrubbed' instead of continuing to the next stage.

Accidents were common, but at Archerfield, not fatal. Inexperience and undeveloped skills would result in rough landings or collision with other aircraft. As there were about twenty-eight aircraft for approximately thirty students, officers were not happy if an aircraft was rendered unusable.

Drills, training, and classes continued, although focus was now on the next time each man could climb into the cockpit.

For the next month, Fred continued with more intense training, which included manoeuvres and emergency landings. One of the greatest challenges was instrument flying in an aircraft built well before radar was developed. Fred would sit in the front cockpit with a hood over his head while Reg executed unexpected turns and dives. On the map sitting on his knees, Fred plotted what he felt or believed the direction in which they were flying. Without the sense of sight, equilibrium and spatial geography were greatly compromised.

At last he could do two cross-country flights - the first as co-pilot, and then, joy of joys, solo with Reg as co-pilot. His final flight test was supervised by Flight Lieutenant Wedgewood, who was very business-like and terse. Apart from barking commands of various manoeuvres, he said little and made no comment on landing. Fred had to sweat it out for his results. By July 17th, Fred had fifty flying hours, half of which was solo.

The flight test marked the end of Elementary Training. On July 18th postings were listed on the wall of the mess. Twenty-five L.A.C.'s, (Leading Aircraft Cadet) including Fred, were being posted to Bradfield Park, Sydney, prior to transport for further overseas training. They did not know where, until close to embarkation.

Fred clapped his hand on the shoulder of his new friend, Charlie. "Did you see that! We're really going overseas! I wonder where!"

Later, he retrieved his logbook from the chief air instructor and tentatively opened it to the fourth page. Officially stamped and signed, a comment read: *Pilot: Average. Judgement and co-ordination weak and instrument flying below average.*

Granted five days embarkation leave, Fred returned to Toowoomba for what would be the last time for four years.

"Well, it has happened - I've been posted overseas, sometime after the 24th. Mother, please don't cry - I'll be back for Easter, at the latest."

Fred spent several days visiting friends. His mate Kev was wistful and envious.

"Lucky bastard, Fred! I'll see you when you get to England, I'm sure."

This time at the station, a sombre mood was hanging over the platform, as families tearfully farewelled sons and lovers. Fred's family was no different. Clara was already pulling her second hanky out of her handbag. Heather was sniffling, and Kathleen would not stop hugging her brother. Finally, Fred looked at his father and extended his hand. This time, the handshake was returned. Bill's face registered signs of resignation, pride, and sadness. To Fred's surprise, he saw a glimpse of tears. Clearing the lump in his own throat, Fred picked up his kit, ready to board.

"Look after one another. I will stay safe, and I will come home. I'll write when I can."

Then he was gone. The train slowly chuffed out of the station. No one moved until it finally disappeared out of sight.

"He's gone! Really gone!" sobbed Clara. "We may never see him again!"

"Cut it out! Bill retorted.

Little did Clara know then, but she would outlive her son by over seven years.

Fred, with other departing servicemen, arrived by several train changes at Bradfield Park, Sydney. Here, he reported to the embarkation depot for final medical examinations and upgrading of kit and uniform. A few days before embarkation, he had some free time to explore Sydney with his camera in hand. Aware that his life was about to change radically, Fred felt a strange sense of false normality. The day before sailing, all men were kept in barracks, having learned their destinations.

Fred was bound for Canada.

Six

Across the Pacific

8th August 1941

Fred and three hundred other servicemen embarked on TSS Awatea, a civilian passenger ship. There were also about two hundred civilian passengers, so these men destined for future hardship had one last trip in relative luxury. This was the sixth and last voyage between Sydney and Vancouver. HMAS Sydney escorted her until she was almost in American waters.

When TSS Awatea docked in Vancouver, she was to be requisitioned by the British (UK) Ministry of War for conversion to a troop ship, and posted to England, for transports to North Africa. Built in New Zealand in 1936, at a length of 527 Feet, and powered by two large turbine engines, she was known as "The

Greyhound of The Tasman" because of her ability to sail at high speed. I am convinced that Dad had an angel on his shoulder, as The Awatea was bombed off North Africa, just over a year later. Fortunately most troops had already disembarked, and no fatalities occurred.

In a letter home, Fred wrote,

> Well, I am at sea, but in comfort, as this ship is still rigged for civilian passengers. I am sharing a second-class cabin with a couple of chaps from my course. The saloons are very inviting and offer entertainment for the non-military passengers. At times we could easily forget why we are here, but we are reminded every day that this is not a holiday - we have to continue our study, attend lectures, and do emergency drills twice daily.
>
> The sea has been very rough at times, but my stomach has held up so far. Thankfully, the food is good, and like what we have at home. We stopped at a couple of ports, but I can't say where. At least we could stretch our legs and look around. By the time you get this letter, I will already be somewhere else. Interesting life so far.
> Your son Fred.

Two days after leaving Sydney, the ship docked in Auckland for three days, so Fred and his new mates caught a train and visited Rotorua. After billeting with a very friendly family, Fred and his friend Charlie visited the sulphur hot springs. That evening, they were entranced by a Maori concert.

Returning to Auckland the next day, they reported 'back on board', ready to sail for Suva, Fiji. Writing on a postcard to send home much later, Fred commented,

> *Bloody hot, and a bit primitive, but the locals are very friendly - plenty of fresh fruit.*

After Suva, came the long voyage across the Pacific with all its moods. Servicemen and civilian passengers mixed at mealtimes, exchanging stories of Australia and New Zealand.

Despite the seemingly normal atmosphere on board, no one forgot what was happening in the world outside. Most of the recruits wanted to talk about anything but the War but there was a deep curiosity from the others, most of whom already had family serving somewhere overseas.

Severe storms tested the Awatea, which had a reputation for pitching and rolling in rough weather. She did not disappoint. The dining room was almost empty over a few days, while the deck railing was overcrowded! Thankfully, TSS Awatea sailed under the Lions Gate Bridge into dock at Vancouver in the first week of September.

"Well," thought Fred to himself as he and the other recruits walked down the gang plank, "This seems more like a holiday than a war, just now."

A transport bus was waiting to take the recruits to billets for one night before boarding the Canada Troop Train. Over five days, hopeful airmen would travel through the Rocky Mountains eastward to several training camps. Wasting no time, Fred, once he was established at his digs for the night, took the map and bus timetable given to him by his hosts, and went sightseeing. What a beautiful, clean city - and all that water!

On a postcard to be mailed much later, he wrote,

> *Here I am on the other side of the world, in a beautiful city. I am seeing places and people I would never have been able to meet if I had not joined up. Am well, and safe.*
> *Your son Fred*

The troop train was not luxurious, but adequate, with seats which converted to bunks at night. A canteen car provided basic meals in between stops.

The route from Vancouver was a tourist's delight: Jasper, Edmonton, Saskatoon, Winnipeg, and finally south-east to Toronto. Fred and his comrades were enthralled by snow-covered mountains, and pristine lakes.

On the back of a postcard illustrating Jasper, Fred wrote:

Mother, if I ever get the chance, I will bring you to this beautiful place after the War. You will need your gloves, and wool socks, though!

Each time the train stopped to load more coal, water and supplies, the men had several hours to explore the towns. The locals gave them a hero's welcome.

"I haven't done anything, yet!" thought Fred, sensing that his 'holiday' was very temporary, and near its end. He bought a large collection of postcards, including places he most likely would not see. Lake Louise, Calgary. A feminine hand had later written neat notes on the back of photos from Calgary. Who was she? Did she have them ready to hand to the servicemen? Perhaps she worked on the catering car.

From Toronto, transport trucks would take the virgin pilots to their posted training bases. Fred found himself travelling to Camp Borden, a Canadian Air Force base, which would be home for six months. Autumn was already showing off with leaves turning to orange and gold on the maple trees. Winter would be a challenge for these Aussies and New Zealanders, who saw it as all part of the adventure, at least for now. Camp Borden marked the change of these young men from adventurer, tourist, and would-be heroes, to skilled airmen, ready to sail to yet another foreign shore where danger was ever present.

Seven

Canada

September 1941
Dear Mother and Dad,
Well, I am in one place again for a while- for about four months, we have been told. I am in a beautiful part of the world, and safe. At this stage, we are far away from the war zone, as we do our next training course. This will decide if I get my wings, and finally help finish the war. The news seems to get worse every day. Our base is quite good, with Nissan sheds for barracks. As the weather can be very cold in winter, wood heaters are at either end of the huts. They also come in handy for boiling water for tea or coffee. Sometimes we have to fight for space around them. It was quite warm when we arrived, but now it is raining a lot, and temperatures are dropping. Snow will be the next thing. That will be a challenge for us Aussies. This course has a mixture of pilots from all over

the Commonwealth. There's about seventy of us, all needing to graduate, so that we can continue in the air force. The scenery here is quite different from home - mountains and streams. When we get a chance, we like to catch a bus into the local town, where we are made welcome. Sometimes, it is nice just to take a stroll if the weather is good. I was given a brochure about the local community. It gave us information about what entertainment was on in town, plus a free bus to get there. On Sundays, there is a concert in camp, at a small cost. It breaks up the tedium of military life and gives everyone a chance to mix outside the cockpit. More news soon.

Your son Fred

In 1916 The Borden Military Camp, named after the former Minister Militia, Sir Frederick Borden, was established by the citizens of the town of Barrie, on a glacial moraine. providing both army and flight training. Between the two World Wars, the aerodrome was used as an acrobatic flying school. With the advent of World War II, the area was again subdivided into two camps for army and airforce training. The Royal Canadian Air Force implemented the Empire Air Training Scheme for Commonwealth aviation recruits.

These schools were the most important training facilities in Canada during World War II. Of prime importance was the fact that airspace was clear and, more significantly, Canada was not in the war zone. Servicemen were able to train in relative safety.

Fred's first flying instructor was Flight Officer, later Flight Lieutenant, Don Bissell Gardiner, a Canadian who had graduated from Initial Course No 4 at Edmonton, April 1940, approximately one year before Fred began his training in Brisbane. Don was the same age as Fred, and was acting as senior officer when he was killed while piloting a Harvard aircraft, which crashed on night training at nearby Edenvale in February 1942. He was only twenty-three and never saw combat service. Tragically, such accidents were relatively common.

While at the Royal Canadian Airforce Camp Borden, hopeful pilots would develop more skills which would enable them to train on combat aircraft in England. Unlike the single-engine Tiger Moth in Australia, the aircraft of choice were sleeker trainers named Yale, after the U.S. university. The longer length provided better centre of gravity. It was especially designed as an intermediate teaching tool, before airmen graduated to faster aircraft.

Fred flew with Don Gardiner in a twin-cockpit Yale, which looked streamlined in comparison to the Tiger Moth. The same arrangement of the pilot in the rear cockpit continued, but this aircraft had the relative luxury of a sliding overhead canopy.

On 15th September 1941, Don as pilot and Fred as passenger, took off in Yale 3450. It was one of 120 Yales provided for training. This flight was labelled 'familiarisation'. The next day involved classroom training on the aerodynamics of a Yale.

17th September 1941

"Right, Fred, you have the controls - get the feel of it. Nothing fancy - just get the feel of her then I'll show you how she stalls."

Fred enjoyed the feeling of flying again, but obediently relinquished the controls back to Don when ordered. Don cut back on the throttle and let the aircraft stall. Almost at the point of no return, he opened the throttle again, pulled the stick back and the aircraft stopped its downward dive and resumed straight and level flying. Learning how to cope when things go wrong was crucial for pilot and aircraft safety, especially in combat. During the next two weeks, Fred continued as co-pilot with Don, practising the controls while airborne. Finally, Fred flew solo, with Don as co-pilot, executing 'medium turns', and 'powered take-offs and landings'. Those manoeuvres, among many others, would keep him alive over the next few years.

4th October 1941

Fred wrote in his logbook:

Certified that I have been instructed in and fully understand the cockpit, fuel system, oil system, performance of the engine, performance of the aircraft, and all ancillary controls of N.A.64 P2 Yale Aircraft.

The following day, Don took Fred 'up' in the next aircraft which would hone his airman skills even further - the Harvard, also named after a U. S. A. university. This aircraft was more complex, and much faster than the Yale, so would challenge the skills of the aircraftmen. Although Fred was a passenger on this first flight, he found the experience thrilling and exhilarating.

That evening in the mess, Ken, a fellow Australian, was curious about Fred's experience. "How'd it go in the Harvard? I was watching you, wishing it was my turn."

"Bloody good, mate! She's got some guts under the hood! Can't wait to take her up solo!"

Ken nodded enthusiastically in agreement.

For Fred, flying was addictive, and becoming more so. That feeling never left him. In a non-combat area, despite the seriousness of the reasons why he was training, he was revelling in the experience. For three days, Fred flew with Don, practising the same manoeuvres he had learned in the Yale. To his disappointment, he then was ordered back to the Yale, for more solo practice.

17th October, 1941

A week later, he flew his first solo in the Harvard. Fred wrote in his logbook:

This is to certify that I fully understand the fuel system, endurance data, hydraulic system, and ancillary controls of the Harvard MK II Aircraft.

This approval signified that Fred would continue with his training. Not all men were so lucky. They were posted to other roles, such as wireless operator.

All through October and November, Fred practised in both aircraft as co-pilot, and more frequently, as pilot. In the Harvard, he learned how to put the aircraft into a spin and then to return to straight and level flying - whew! Fred's stomach churned, but his eyes glistened with excitement. He would need that skill when avoiding enemy aircraft over Europe.

After one flying exercise, Fred saw Ken, with bucket and rags, looking very green.

"What's wrong, mate?"

"I have to clean up the cockpit - I spewed everywhere! I feel like a mug!"

"I'll buy you a beer when you're finished, mate," joked Fred.

Ken groaned and took off for the latrine.

15th November 1941

Fully qualified to act as a safety pilot.

(Fred was now qualified to maintain visual separation from other planes, clouds and terrains while supervising trainee pilots who wore vision blocking devices for the purpose of learning instrument flying)

8th November 1941

<u>Wings Parade, Saturday Evening, Drill Hall</u>

Heavy snow had hampered some flying assessments, so not all airmen had yet had their final wings tests before graduation. They were judged on their previous practice and theory exams. The night of the graduation was affected by snow as well, so instead of a presentation in the parade ground, it had to be conducted indoors.

Group Captain R.S. Grandy, O.B.E., addressed the graduates:

"You have reached the stage in your air force career when you have qualified for your wings, and for the opportunity of receiving advanced training at an operation training unit. The personnel of both flying squadrons have had to operate under a handicap during recent weeks. The fact that this Class 36 of No 1 Service Training School has been able to graduate on this date reflects great credit on the part of all concerned. The extra effort put forth by all ranks is commendable indeed. I am glad to be able to say certain promotions have come through today. The names of those promoted will be announced by the adjutant. I wish you Godspeed and good luck to you all."

<u>(courtesy of No 1 Service Training School Camp Borden Blog)</u>

Because training and wings tests were unfinished, celebrations were postponed for two weeks. When the weather finally cleared, the graduates had the opportunity to practise formation flying, which was much more dangerous than sharing the sky with only a few aircraft. A formation of twenty or more demanded absolute concentration, and awareness of the positions of the other aircraft.

25th November 1941

Fred raced off to the latrine for the third time in an hour.

"What's wrong mate? Are you crook?"

"No - nervous - I've got the runs - I'm doing my wings test in an hour."

"You'll be right mate, you've already graduated!"

"I know, but they might change their minds."

Fred passed his wings test, but practice continued with night flying, instrument flying, and then, ultimately a mock battle where all pilots raced to their aircrafts, took to the skies, where they dived and turned, practising mock attacks and defence manoeuvres. All too soon, they would be experiencing the real thing.

5th December 1941

Fred flew the Harvard for the last time in Canada, clocking up a grand total of 140 flying hours.

19th December 1941

The base commander wrote in Fred's log book:

*LAC Bright (Lance Air Cadet) was **not** given his wings navigation test due to bad weather. He is assessed on his solo exercises.*

Final results:
As a Single Engine Pilot - Average
As a Pilot-Navigator - Average
Needs more practice with instrument flying/compass flying

Fred was also promoted to the rank of Warrant Officer.
The dangers of flying became all too evident in the years after Camp Borden.

Fred was one of six Australians who graduated. Only two of these men flew through the war, relatively unscathed.

Fred succeeded in avoiding death and injury, except for gashing his leg on a kerosene tin while in North Africa, requiring stitches and six weeks off flying. One officer became Squadron Leader of Squadron 453 in England. He was awarded a DFC. (Distinguished Flying Cross.)

Tragically, the others did not see home again. One airman was shot down over Germany. Another was accidentally killed in New Guinea, in 1944. A third airman was shot down over France, and declared Missing in Action, Presumed Killed and buried in France. The last of the six was killed in a reconnaissance flight crash over New South Wales, Australia in 1942. A small compensation may have been that he saw his loved ones before his demise.

As the next wave of pilots in training was due to arrive, the graduated pilots relocated to a base camp in Toronto. At last, they had some leave, and the chance to do some sight-seeing. Ken and Fred visited Niagara Falls, which were still flowing, despite the extreme cold.

"Bloody amazing!" shouted Ken.

Over the roar of the falls, Fred shouted, "I can't bloody hear you!"

Fred enjoyed using his camera again - he was fortunate enough to see a grumpy bear, which should have been in hibernation by now.

Travelling back to Toronto, a few of the Aussies explored this beautiful city, which had some of the shops below ground, for protection against the winter cold.

Fred and Ken decided to take the train from Toronto to New York for Christmas. The THB, (Toronto-Hamilton-Buffalo) had a through service to New York. Fred and Ken booked the No 74, a café Parlour Car, which meant that they would be sitting in their seats for the journey. The journey was long, estimated to take more than fourteen hours travelling at an average of forty-five miles per hour. The rail line passed by Lake Ontario and Lake Erie, before crossing the U.S. border into Pennsylvania, then New York State and on to New York City. Arriving at Penn Station on 34th Street, they found a cheap hotel in Mid-town Manhattan and walked to Times Square.

Compared to Australia at the time, Manhattan with its lights and advertising signs, was magical. As servicemen in uniform, the young men were well treated by the New Yorkers. Rudi, an Italian restaurant owner, kept the bar open for Fred and Ken, after all the diners had left.

Later, Fred sent a self-timed photo of Ken and himself, seated at a table with a half-empty bottle of whisky, and cigarettes. He wrote on the back,
"*Having a few quiet drinks – Merry Christmas!*"

They enjoyed spending their accumulated pay by exploring bars and restaurants, and going to the Ethel Barrymore Theatre, to see *Best Foot Forward*. A bit of charm and flirting with the box-office girl ensured that they had good seats! A uniform can be very alluring for some.

As they had leave until January 2nd, 1942, Fred and Ken decided that they may as well stay on in New York to watch the ball drop in Times Square at Midnight on New Year's Eve.

On New Year's Day, suffering from no sleep and bad hangovers, Fred and Ken returned to Toronto on the THB.

The postings were out. The Australians were told that they would get the train to Halifax, ready for embarkation for England on January 8th.

In the Halifax harbour was anchored the Convoy HX 169 fleet, consisting of forty-two merchant ships and sixteen escort ships.

The merchant ships carried cargo and a mixture of passengers, civilian and military. Fred and Ken fortunately were allocated the same ship, The Beaverhill. Built in 1928 at 500 feet in length, she was a rather elderly ship, with the promise of an uncomfortable voyage to Liverpool. The weight load consisted mostly of essential goods, but it had room for sixty passengers.

Although she survived combat, The Beaverhill later ran aground in November 1944 off St Johns Harbour New Brunswick in rough weather, and had to be scrapped.

On this day, Fred realised his ten months of adventure and sightseeing were over. In stark comparison to his voyage outwards to Canada, Fred found himself below the water line, sharing space with twenty other passengers, sleeping in hammocks. Under the hammocks were the mess tables. Every morning the hammocks would be hoisted to the top of the cabin space to allow the tables to be used for meals.

The crossing of the Atlantic can be rough at the best of times but in this 'bloody old tub', as the servicemen called it, made life miserable. What food was available, if eaten, did not stay down for long. Getting fresh air on deck was a luxury to be enjoyed when no enemy vessels were detected on the ship's radar. Fred and his fellow passengers spent days on end with heads over buckets.

"I don't care if I die!" was a common refrain from the travellers.

Blessedly, calmer weather arrived, and the surrounding seas declared devoid of the enemy. Fred sprawled on the deck, flat on his back, sucking in fresh sea air. His trousers were loose after days of seasickness.

Of greater peril were the enemy submarines and torpedo ships seeking allied vessels. The escort ships formed a perimeter around the merchant ships, with some diverting for other ports, and others joining the convoy. HX 169 was lucky, with no attacks recorded.

The reality of war was descending on the new pilots.

At last, the coast of western England showed on the horizon. On 23rd January 1942, the convoy docked at Liverpool, in the cold and rain of a different northern winter.

On disembarking, several of the men knelt down and kissed the ground.

"What have I done!" muttered Fred, as he and the rest of the military men onboard were taken by bus into Liverpool, in all its dreariness. A short stay in this city, and Fred would be on the train to Bournemouth.

Fred, Canada group photo

Fred, flight training, Canada

Fred and Ken, New York. "Having a few quiet drinks - Merry Christmas"

Fred and Ken, New York

Eight

Reality Approaches

23rd January 1942

On a cold, wet and miserable Midlands morning, Fred waited on the Liverpool station platform in the company of a large throng of servicemen and local civilians.

"I hope they're not all going to Bournemouth," shouted Fred, to Ken, over the noisy crowd.

A British soldier, overhearing the remark, shouted back, "No, chum, this is a central station for servicemen being posted, and going on leave. Everyone is everywhere, like ants!"

"Thanks mate!"

Eventually the London and North-western Railways locomotive pulling twenty passenger carriages, steamed into the station. Fred and Ken, kits over their shoulders, pushed their way onto No 5 carriage. Thankfully, they found a couple of seats, and after stowing their kit on the overhead luggage rack, collapsed onto them, taking possession.

"Where am I supposed to bloody sit!" complained some late-comers.

"On the bloody floor!"

The windows of the carriage were already misting over with condensation from the fog of breath, some sweet and some foul, in a confined space. A loud toot and the belching of smoke announced departure. Two hundred and fifty-six miles would feel like another ocean voyage by the time this journey ended. Fred felt a tightening in his stomach, as he wiped the carriage window and looked out at anonymous towns with their rows and rows of terrace houses. Names had been removed, to protect identity from the enemy. As night closed in, the houses, with windows blacked out, became invisible. This left a sense of isolation and bleakness with the fatigued travellers.

Seven hours later they pulled into Waterloo Station on a dark, cold night. Grabbing their belongings, everyone scrambled along the platform, up a set of stairs, across a pedestrian bridge, ready to board the London and Southwest train, bound for

Bournemouth via Southampton, interrupted by several stops at unnamed towns.

Those stops gave the servicemen the chance to grab a cup of tea in a cold station waiting room. Coal was scarce and was being conserved for essential services such as the trains.

"As cold as a witch's breath!" muttered Ken, cupping his hands over a tin mug of white tea.

Returning to their carriage, Fred and Ken found two R.A.F servicemen in their seats.

"Oy! Get off there! We were there first! We are senior to you!"

Grudgingly, observing the wings on the Aussies' chests, the R.A.F fellows obeyed with, "Bloody colonials!"

"Bloody poms!"

As a couple of Military Policemen were at the end of the carriage, all parties decided to let things be.

24th January 1942

"We must be getting close, boys!" Fred shouted. "I just saw the ocean!"

The train stopped at Southampton, where the men were allowed to disembark, and find somewhere for breakfast.

"And a toilet," thought Fred, who was far from excited about the one foul toilet at the end of his carriage.

By lunchtime the trainload disembarked at No 3 Personnel Depot, Bournemouth.

In a letter home, Fred told his family about the latest leg of his ongoing journey:

> Dear Mother and Dad
> I have been on the move again, after a rough trip across the ocean, then one day and a night on a train. Conditions here are poor after our last digs. We are being put up in some old pubs that should have been knocked down years ago. The menu of war rations is not very appetising, but at least it is food. I can't say much more for now, but I am keeping safe, while waiting for my next posting. Hope you all had a good Christmas. Your parcel finally caught up with me - thanks for the boiled fruit cake. It lasted well. I had to fight the other lads off! When I know my next move, I will let you know. Sorry that I can't tell you where I am, but the enemy has ears and eyes everywhere.
> Love from your son
> Fred

<u>No 3 Personnel Depot at Bournemouth, on the English Channel.</u>

Formerly a pre-war holiday resort, the town was now a sorting base for incoming servicemen from overseas. Apart from about ten Australians who were billeted in a run-down hotel, several Canadians stayed in a second, decrepit pub. The buildings were cold, dank, and musty. Beds were lumpy and uncomfortable. On good days, the toilets worked.

Fred knew that he was now at War. From across the Channel, the Luftwaffe would descend over the coast in air raids. It took a while to get used to the sirens when all personnel took shelter wherever they could. Fred would hold his breath, listening for aircraft and the whistle of falling bombs. Sometimes, he breathed a sigh of relief when the sirens sounded an "all clear" after a false alarm.

The Australians, used to walking on their own nation's golden sandy beaches, were prevented from going onto the rocky beaches at Bournemouth, as it was completely closed off with barbed wire - out of bounds.

Following (another) medical examination, and new kit issue, the newcomers were granted leave. They wasted no time taking an available train to London. Although London was sandbagged, and draped in blackout curtains, the servicemen were relieved to have a change of scenery, and to visit the city of their ancestors. Fred's great-grandfather had been a shopkeeper outside London.

Fred felt a connection with his roots, and a great sense of patriotism. If he had the time, he would have liked to seek out the family shop. He would be fighting for the Empire, as well as the country of his birth.

28th March 1942

On returning from London, Ken and Fred walked into No 3 Personnel Headquarters, which was one of the slightly better Bournemouth hotels. As they expected, the postings were pinned to the wall at Reception.

"Well, mate, it looks like we'll be going in different directions," commented Ken, who saw that he was being posted to an Operational Training Unit in Ossington Lincolnshire. Fred's destination would be North Africa.

31st March 1942

On a foggy night at Bournemouth, when the possibility of an air raid was unlikely, Fred and Ken went to the local pub for sausages and mash, and a pint or two.

"Well, it has been good having you for company these past few months," said Ken.

"Yes, mate - it keeps the homesickness at bay."

"You, Fred Bright, homesick! First time I knew! I thought you were tough!"

"Not as tough as I look, Ken. I learnt to act that way a long time ago. It helps when I'm not tall, like you."

Ken pondered, "I wonder what will happen next."

"I guess we just follow orders, and get our crews soon, I hope. Bit more training first though, I expect. Whatever happens Ken, just keep yer bloody head down!"

1st April 1942

After an early breakfast of powdered eggs and chicory coffee, Ken and Fred parted company, with a customary handshake, slap on the back, and clearing of throats. Ken and several other aircrew personnel boarded a train for London, then on to York, and Ossington.

Fred went by a transport jeep to St Eval Airfield, about one hundred and sixty miles west of Bournemouth. Normally, he would have been ordered to travel seven hours by train, but as he was to meet with a Wellington ferry crew ready for departure, he had the luxury of a four-hour drive instead. That trip would be the last of some relatively comfortable travelling for a while.

The Wellington bombers usually had a crew of six. As Fred was a passenger/observer, only one wireless operator boarded the aircraft, instead of the usual two.

Harry, the British pilot, looked at Fred's kit in annoyance. "We're not a ship, you know! That kit will be your seat for this trip. Take this 'chute'- we'll be flying over enemy waters."

For once, Fred felt unsure of himself. He threw his kit into the belly of the Wellington, and then hauled himself up through the hatch. The gunners were already in place, secured into their hammock seats. Charlie, the navigator, had a small fold down stool and desk behind the cockpit. Pete, the radio operator, had a similar set-up on the opposite side.

"You'd better park yourself over here, where the other WOP (wireless operator) usually sits. How's your navigating?"

"Not bad," lied Fred.

"Good - you might give Charlie a break later - he'll need it. We are going on a mystery tour to keep Jerry off our tails."

Harry was on orders to follow the coastlines of Spain and Portugal, en route to Gibraltar, a British territory at the tip of Spain. Fred clenched his buttocks, fighting off the urge to shit himself at the thought.

The Wellington, known as The Wimpy, or Widowmaker, did not have a good reputation, with its lack of mechanical reliability. Although it was faster than its predecessors, it was difficult to fly and manoeuvre.

Pre-flight checks completed, Harry started each of the two engines in turn. At the 'clear for take-off' radio call, the Wellington rumbled forward gaining speed, finally lumbering into the air. This aircraft was built for defence and speed, not comfort. The pilot had the only seat which provided a semblance of ease. Each man had filled a flask with soup before take-off. This would be their only sustenance until Gibraltar.

By leaning forward, Fred managed to have a short conversation with Pete.

"Is this your first trip?" asked Fred.

"No, second. I'm more nervous this time, after nearly ditching into the sea last time. You?"

"This is my first time flying in enemy airspace. Bit of a change from the past year or so. Guess I had better get used to it."

An hour later, Charlie leaned back on his stool, and signalled to Fred. "Can you take over here for a few minutes? I need a leak. We're not far off the Portuguese coast, so you should be O.K.."

Fred took Charlie's place, and compared the marks on the map, with the image of the coastline with what he could see through the small window above the navigator's desk.

On his return, Charlie checked position and nodded with satisfaction that the aircraft was still on track. Charlie relied entirely on his charts for navigation, as any radio communication between crew members would be intercepted by the enemy.

Nine hours after leaving St Eval, Harry successfully landed The Wimpy on the Gibraltar airstrip. Fred, as soon as he could deplane, raced for the latrine - blessed relief! That Wimpy had soared, ducked, and dived, while dodging enemy aircraft. For Fred, this would soon be all in a day's work.

Gibraltar played a significant part in the theatre of war. R.A.F North Front was opened in 1942, later to be declared Mediterranean Air Command. Fred would, in due course, become very familiar with this part of the world. Intelligence had warned Gibraltar that Malta, the next stop-over, was being bombed, so the crew had a two-day break in relative peace.

Although most of the civilians had been evacuated to England, a few shopkeepers remained, servicing the increasing numbers of military personnel. Fred wandered from his billet to a small shop which sold postcards featuring one of the several families of monkeys for which Gibraltar was famed. On Europa Point, he found a seat, and wrote on the postcard he bought.

Dear Mother and Dad

I am on the move again. Life is not dull. Weather is warm, and I am safe. Hope all is well at home. Letter coming soon.

Fred

Charlie joined Fred on the seat. "Wondered where you got to."

At that moment, a small monkey sprang on to Charlie's head, chattering loudly. "Shit! What was that!"

Roaring with laughter, Fred replied, "Someone's making a monkey out of you!"

Fred grabbed his camera and photographed a startled Charlie and a cheeky monkey.

At that point, Harry arrived. "Right oh, chaps! We've got clearance to leave for Malta tomorrow. Time for a pint, and an early night."

At 0600 hours the following day, the crew flew towards the island of Malta, south of Sicily.

Aircraft on the tarmac were regularly blown up on the runway. Harry flew low over the Mediterranean to avoid attention from the enemy in nearby Sicily. Just over eight hours later, the group landed safely to the great relief of all. Taxiing was intricate as there was a line-up of other aircraft either having maintenance or waiting for a break in the bombing to take-off for its next point.

An air raid siren wailed while the crew was in the Mess having dinner. Military and civilians filed as quietly as nerves allowed, to the bomb shelters and waited. Over the next three sleepless nights, life was disrupted by constant trips to the air raid shelter. Aircraft maintenance was delayed. Some planes were destroyed.

During the second night in the shelter, Fred found himself sitting on the ground next to a small girl with huge brown eyes, staring curiously at him. Thinking suddenly of his younger sister Heather at home, Fred smiled at the girl. Bombs were whistling down outside. Fred resisted the urge to duck.

"Want to see a trick?" he asked, not sure if she understood him.

The girl nodded. Fred pulled a piece of calico out of a pocket. He always kept it - it came in handy for wiping his hands when frequently no sinks were available. Fortunately, this piece was clean and square. Slowly and carefully, he folded it in half diagonally, repeated the process, and finally folded up the edges, to form a small calico boat.

"Oh!" exclaimed the girl, smiling now, despite the air raid raging outside.

A moment's normality seemed almost like a miracle.

"You can keep it!"

With a second smile, she took the calico boat, and slid sideways towards her mother, who had been looking on quietly. As she nodded and smiled, the all-clear sounded. Military and civilians slowly filed out of the shelter, hoping that there would still be some sleep for the remainder of the night.

7th April 1942

Ideally a relief transport crew would be on call for the next leg to North Africa, but that crew had been shot down off Sicily. There was no choice but for Harry and crew to fly on to Cairo when there was a window of opportunity.

Finally, Harry's Wimpy was at the head of the queue for departure. Radar and wireless operations had detected no enemy aircraft nearby for the past hour, so Control gave the go-ahead for take-off. Six hours later, the Wellington circled over Cairo, which, while also being menaced by the enemy, was tranquil in comparison to Malta. Finally, the ground rose to meet the landing wheels.

Although a British territory in 1942, Egypt was home for an ancient royal family, currently headed by the difficult King Farouk, who ignored night-time blackouts by having all the candles illuminating his palace at night. Of greater importance was the fact that he had vehemently resisted British military intrusion, even though the Allies had rescued Cairo from an Italian invasion. Now the Germans were a much bigger threat. Following an allied military ultimatum, King Farouk capitulated, allowing a more orderly government. Consequently, not all Egyptians were enticed by the British and allied military.

Fred was relieved to arrive, but was sad that he would be parting with this courageous crew who, after having a few days leave, would ferry another aircraft to India.

Fortunately, Fred also had five days leave before travelling to his unit. Harry had mentioned a servicemen's houseboat on The Nile - dubbed the *Araba*. It provided comfortable accommodation, meals and recreation for military personnel taking a break from the ardour of war.

In the letter promised on the previously sent postcard, Fred wrote,

> *Dear Mother and Dad,*
> *I am now almost where I am meant to be for the next few months. At last, I have had a good bath and a few decent meals. I am enjoying the company of a few English lads who are quite decent in spite of my earlier impressions. We have been busy sight-seeing an ancient city, where beggars and children chase us constantly, asking for money. We had our shoes polished for a bargain! I am enjoying using my camera again, taking photos*

of old buildings and churches. Some of the locals don't like being photographed, so I have to be careful.

After the last ten days, it is nice to be in one place, at least for a while. Tonight, there will be a tea dance at my digs. Haven't seen women in uniform for a while, so that will be a pleasant change. I'm not sure exactly where I'll be next until I get there but will try to write again soon.
Love to all
Your son Fred.

15th April 1942

"Thanks chaps for your company over the past few days. Thanks for the lift. Keep your heads down!"

Fred gave the Wellington crew a final wave as they left by jeep for the airstrip. He had one last day before a R.A.F transport truck would collect him and six other R.A.F and R.A.A.F airmen to take them to a maintenance unit, where preparations for a new base would be started. These past few days had a sobering effect on him. The vision of the little girl in the air raid tunnel in Malta had not left him. Why did innocent children have to get caught up in man-made wars? With each passing day, Fred realised that he would willingly put his life at risk to protect the vulnerable in the Middle East and also his family at home. The sense of adventure had turned to commitment and growing courage.

Nine

In the Desert

April 1942

<u>Abu Sueir Airfield, C & M.U (Crew and Maintenance Unit</u>, seventy-two miles from Cairo. It provided strategic protection for the Suez Canal.)

Fred's war finally began. He was about to get a fundamental history lesson. During the early 1940s Mussolini invaded parts of the Middle East, bombarding it with land and air raids. For a year, the Allied Forces had succeeded in driving back the Italian forces on the ground and in the air. Hitler then started pushing his forces forward, resulting in devastating attacks on the Island of Cyprus, with huge civilian loss. Rommel's troops had entrenched themselves in the Middle East.

In response, the R.A.F planned to introduce long-range heavy bombers. All hands were required to restore the many damaged airstrips and prepare for the arrival of Halifaxes and Wellingtons.

Fred found himself 'somewhere in the desert' with no plane to fly. He could only guess that he was one day's drive from Abu Sueir. His accommodation consisted of a tent in a city of two-man tents that offered a mess tent and an open-air latrine. With other aircraftmen, he was directed to help the groundwork crews with repairing the damaged runways by helping to lay down Marston Mats (planks of perforated steel) which would form a solid landing platform for heavy aircraft. The perforations lightened the weight of each sheet. They were laid down in a zig-zag pattern similar to modern floor matting. An engineer periodically interrupted the work crew by checking for booby traps left by the retreating enemy. Anticipating an explosion, the work gang quickly retreated to a safe distance until the all-clear whistle blew, when everyone would exhale with relief.

After a day of helping to lay ten feet long, fifteen inch wide sixty-six pound steel sheets on to what would become an extended single runway of 1000 feet, Fred collapsed on his stretcher in the tent he shared with British navigator, Tom.

"Bloody sand and flies!" Fred complained.

"Thought you would be used to it."

"Flies, yes - sand at home is mostly at the beach - a much nicer way to have a holiday."

Fred now wore tropical style uniforms - open necked short sleeved khaki shirts, khaki shorts, and long socks in light shoes, for non-flying duties. He became very lean and muscular with the constant physical work and rations of tinned meat, tinned potatoes, and vegetables, as well as tinned milk. Bread dried out too quickly in the desert heat, so service biscuits were supplied as an alternative. Cured bacon was occasionally available, but eggs were only rationed to operational military. The others had to tolerate egg powder. Fred decided that tinned bacon could put him off that treat for life. Meat dishes were embellished with curry or tomato puree. For a change, he could eat tinned sardines.

Coffee and tea were only available until supplies were depleted. Otherwise, hydration was available by rationed water. Every two weeks, the ration truck would arrive from Cairo, usually with smaller and smaller portions for each man. With the exchange of rationed cigarettes and the accumulation of their allowance of one and a half pence per day, a highlight would be the receipt of fresh fish from the Cairo markets.

Attempting to break the tedium of hard physical work and not much else, the servicemen would take turns in hitching a ride with the ration truck to Cairo for three days leave and catch a return ride with another truck. While on leave in Cairo, away from the swarms of flies in the desert, Fred wrote a brief letter home.

> *Dear Mother and Dad,*
>
> *I am enjoying some recreation and a few beers in an exotic location. It makes a big improvement on the hard slog I have been doing with the other blokes. Sadly, I'm not flying at the moment. That will happen soon, I hope. We usually only know what is happening next with very short notice. I am looking at the shops for a few things to send home for Christmas - I know it's a bit early, but the mail is very slow, as you know.*
>
> *Hope your food is better than mine. Tinned bully beef has lost its appeal, after two months in my current location.*
> *Love to all,*
> *Your son Fred.*

2nd July 1942

"Bright, get your gear! You are being posted!"

"Where, sir?"

"You'll know when you get there!"

Without fanfare or fond farewells, Fred climbed into the back of a transport truck in the company of ten other pilots who had also spent the past two months repairing the airstrip.

Life would be a little easier for a while, at the P.A.P (Pilots Air Pool), Khartoum. This location had an airstrip near one single tree with a ghoulish history.

It was named *Gordon's Tree*, because General Charles Gordon was beheaded by the Mahdi during the Siege of Khartoum, in 1885. The Mahdi led a rebel religious group which warred against the Egyptians and planned to overtake the British interests in the Sudan. General Gordon's overconfident approach led to the Mahdi holding siege of the walls of Khartoum, and massacring Gordon and his troops.

July 1942 was slightly less gruesome in this location as pilots waited to be posted 'somewhere-anywhere.' Like the others, Fred had some leave to explore the city of Khartoum or stay on the base, and play cards. His life was a bit more civilized here, as he could sleep in a pilot's hut, and have access to better food.

19th August 1942

The boys were on the move again, as higher command was making up its mind what to do with too many pilots, and not enough planes. This time Fred moved north of Khartoum, to 108 Maintenance Unit at Fayum, just south of Cairo. Once again, the pilots waited for a real posting. Maintenance Units served to repair and service aircraft, as well as receiving incoming combat aircraft from Britain. They also acted as a holding station until the 'brass' decided where seemingly surplus aircrew should be posted.

"This is a bloody waste of time! I joined up to fly!" chorused the pilots.

Four days later - at last! Another posting to Cairo finally gave Fred the chance to fly again, although it was only twice with cross-country flights around the Pyramids in a bi-wing Hawker, known as an Andax. It was very similar to the Tiger Moth - Fred grinned as he soared around the historic monuments. Those flights would help Fred maintain his flying hours. Unfortunately, he would have no further chance to fly for some months. Late October, he was transferred to the Pilot Transfer Centre at Alexandria, awaiting a ship to transport many of the aircrew back to Liverpool.

While politicians and commanders bickered about the African Campaign, the servicemen had twiddled their thumbs in harsh conditions, with little sense of satisfaction at contributing to the war effort. Surely that situation would change - some time!

21st October 1942

No 21 Personnel Transit Centre Kasafreet, Egypt.

In a transport truck with Allied airmen for company, Fred travelled north from Fayum to Kasafreet which was home for the 107 Maintenance Unit, and a holding area for servicemen arriving and exiting Egypt, via the Suez Canal.

Kasafreet was a fascinating, harsh town, where servicemen mingled with kaftan-draped Egyptians. Beggars lingered outside the YMCA, which provided billet accommodation for the fortunate few. Others were relegated to groups of tents, each group numbered to identify who was where. Photography was forbidden, so Fred had to occupy himself by wandering through the market stalls, savouring mysterious and aromatic spices. Perhaps this is where he developed a love of curries?

"The hotter the better - it has to make you sweat," he'd always say.

In a small café, he sipped his beer and wrote home.

> Dear Mother and Dad
> I have moved on again, to another hot and dusty place. We are all waiting for transport to our next posting. Of course, I can't tell you, as the censors would cut that out of my letter. There is a mix of British and Australian lads not serving as aircrew here. Some are workimg, while the rest of us entertain ourselves by reminding them regularly that we're not! Late in the afternoon, we play cricket - Poms against the Aussies. The locals are keen to join in too - some of them are not too bad.
>
> I went to one of the cities not far from here. and saw a few W.A.A.Fs off duty. (Women's Auxiliary Air Force, who served by performing map plotting, mechanical and radio duties. They did not serve as aircrew. Author)

I bought a pretty girl a beer, and we had a nice chat. Nice change from these smelly as blokes!

Your birthday parcel was waiting for me when I arrived here - thanks - I see you posted it in May, so it took a while to get here. Thanks for the jumper, Mother, but I don't need it here - a hundred degrees here most days, although the nights can get cool. I expect to be posted to a cooler climate, so I expect that I will enjoy wearing the jumper when I can sneak out of my uniform on leave. I may not be in touch again for quite a while, but don't worry. I am used to all this moving around now, even if it's bloody annoying at times.

In case I don't get a letter to you before, I wish you all a Merry Christmas. The prediction of the war being over by Christmas has not happened. You should get a parcel from me some time in December. Fingers crossed.
Until next time,
Love
Your Son Fred.

1st December 1942

"What a tub!" moaned the military servicemen as they boarded HMT Almanzora, bound for who knew where. This former cruise ship was built in 1916, so comfort was minimal. The NCOs had the pleasure of canvas hammocks for sleeping. A mixture of British, New Zealand and Australian aircrews were sharing confined spaces in oppressive heat. The Almanzora was part of a convoy of twenty merchant ships and escorts. Although the Allies were regaining control of the African Campaign, the route would be long and tortuous, to avoid enemy U-boats and submarines.

One by one, each vessel weighed anchor at Pt Said, and cruised down the Suez Canal, into the Arabian Sea. The early days were quite pleasant, with calm waters until the convoy reached the open ocean, heading south to the southern tip of Africa. The persistent tacking to screen the presence of the convoy, and heavy seas took its toll on those who were dreaming of being airborne. Fred became friends with the bucket again.

"Never again," he moaned. "I just want to die."

The stench of vomit in the putrid cabin set off most of the others. Only at night, could they go up to an unlit deck while they sailed through an area where enemy war ships were detected on the horizon. Respite came at last when the convoy dropped anchor at Pt Elizabeth. Solid ground at last!

Day leave was granted, so the servicemen wasted no time disembarking to find their land legs again. Fred knelt and kissed the ground. He had already been told that this was a stayover, not a posting.

"Bugger!" he thought, feeling sick again at the thought of going back on board.

Pt Elizabeth was busy with Afrikaans and South Africans who seemed to be hovering near the military visitors, grateful to see the arrival of military protection. The servicemen were welcomed warmly, receiving free coffee and biscuits in several cafes.

Fred and a few of his Australian compatriots relished the thought of going to the cinema, at The Grand Theatre. "Mrs Miniver" was on the program. It was a bit of a sob story, but at least a good distraction. Fred looked at his cinema ticket:

"Look at that!" he commented to a new acquaintance by the name of Max. "It says Whites only! Poor buggers!"

The men felt slightly uncomfortable at the obvious sign of racial division, but had to shrug and walk into the cinema. They already had a war to win - the racial conflict would need deep thought at another time.

Two days later the Almanzora joined a smaller convoy which would continue across the south of Africa, through the Cape of Good Hope.

"Cape of no bloody hope!" became the catchcry as sea sickness struck again.

Finally, after three days of pitching and rolling, the Almanzora started up the west coast of Africa, where the seas were choppy rather than tumultuous. Spirits rose, and appetites returned.

Sunday 20<u>th</u> December 1942

Early Christmas Dinner and Crossing the Line.

Menu
Queen's Cream
Fried Fresh Fish, Sauce Tartare
Roast Chicken, Bread Sauce
Cabbage
Fondante Potatoes
Vanilla Blancmange with Fruit
Coffee

In relatively smooth seas which were deemed safe, an atmosphere of revelry permeated the dining saloon. Earlier in the day, "King Neptune" had crossed under a rope at the exact moment as the Almanzora crossed the Equator. Beer and rum were doled out in ration-sized portions, but the alcohol was sufficient to enhance the sense of early Christmas celebration. All thoughts turned to home, wherever that may have been for each man.

10th January 1943

As the small convoy sailed further north, temperatures steadily dropped, until a wintery chill permeated Fred and his comrades. By this time they had been told that their next and final port would be Liverpool, where the airmen would be regrouped for further training in their specialties.

Fred hoped that he would ultimately command a crew. Once again, he had to be patient, and wait.

Fred in Egypt

Ten

From Sand to Fog and Rain

25th January 1943

Nottinghamshire, mid-England. No 14 Advanced Flying Unit

Almost two years since enlisting in the Royal Australian Air Force, Fred advanced closer to being involved in air warfare. Allowing for the age and poor quality of many of the aircraft he had flown previously, the fact that he had not suffered injury or worse during his training to date was a remarkable feat.

In the fourth year of the war, Britain seemed to be struggling against the better prepared German Luftwaffe, thanks to Hitler's manic ambitions from the early 1930's to overtake the western world. In contrast, Britain had anticipated that there would be no war for at least ten years, and as a result didn't scale up air and maritime armaments. Most allied countries did not want to believe that global conflict could be possible, especially so soon after the Great War. However, in 1932 future Prime Minister Stanley Baldwin warned of the risks of air attack on Britain at some time in the future. The Air Admiralty oversaw the production of military aircraft in shadow (secret) factories, where parts were made separately and transported to other areas for final assembly. However, the production rate was much lower than that of World War I, possibly due to the Great Depression, lack of food, and illness. When war was declared in 1939, the Air Admiralty was superseded by the Ministry of Air Production (MAP), which had the authority to source materials and allocated resources for increased aircraft production, as well as using damaged planes for spare parts.

Wellington aircraft started coming off the assembly line from 1940, eventually totalling 11,500 by the end of the War. Lancasters would total 7,500 at war's end. The first Halifax was produced in 1940, totalling over 6000 aircraft by the declaration of peace. This aircraft later required modification for desert conditions, when plans for air defence of the Middle East were made. As the war progressed, MAP (The Ministry of Air Production) adjusted the goal posts to meet the defence requirements. But in the fourth year of the war, Britain seemed to be struggling against the better prepared German Luftwaffe.

No. 14 A.F.U (Advanced Flying Unit) was based in the grounds of Ossington Hall. The Hall itself was originally built in the seventeenth century for the Cartwright family before being passed on to the Denison family in 1768. In 1942, the Denison estate provided the Hall and grounds for R.A.F training. Two concrete airstrips were constructed on the private property. Ossington Hall was used as barracks for the servicemen who were most unhappy with the dank, musty conditions of a two-hundred-year-old mansion. They were also particularly disturbed by the resident ghost!

January in Nottinghamshire was the exact opposite to the North African desert heat. The average temperature was thirty-seven degrees Fahrenheit. Khaki shorts were discarded in favour of dress uniforms, overalls and heavy coats.

Fred rode his new second-hand bicycle through the Victorian iron gates of Ossington Hall, down a tree-lined drive, past a Georgian church to the Hall, where he shared a large, cold room with four other flight sergeants. The fireplace provided small comfort, as it either smoked or went out.

He sat on his bed, with his two issue blankets around his shoulders, attempting to write a short letter home.

Dear Mother and Dad

Thank you for your letters. I have just received them, after being on the move for quite a few weeks. I have travelled from a hot and dusty place to cold and rain - bloody cold! At least my digs are under a roof now, even if the roof does leak. I expect to be 'here' for a while, so will write when I can.

With love from your son Fred.

"Have you seen the noticeboard downstairs?" Fred asked Pete, who was lying on his bed, smoking and reading a *who done it* book.

"Have they posted our courses at last?"

In the grand entrance, a notice listed the training courses about to commence. A group of pilots, including Fred, were to be ferried to Banff in Scotland, to commence training on the Oxford aircraft, prior to learning BAT (Beam Approach Training), a technique developed to enable blind landings at night in bad weather.

The twin-engine Oxford, dubbed the "Oxbox", was specifically designed as a training aircraft for all roles on a bomber aircraft - pilot, bomber, navigator in particular. Although nothing official had been revealed by the hierarchy, all servicemen knew that this posting was the start of hopefully what would culminate in flying either Wellington or Halifax bombers. First, came the training on this 'new' aircraft.

4th February 1943

Fred's first instructor pilot was another "Fred", Fred L. from Adelaide, Australia, the second of three brothers to enlist during the course of the war. He had recently flown from Egypt with No 11 Squadron, completed his missions from Palestine and Egypt during the campaign to regain lost allied territory over Crete. Fred Bright was in awe of being instructed by a man who had already done what *he* wanted to do.

During the following six weeks, Fred learnt and practised flying in the Oxford, which would be a forerunner for multi-engine aircraft. Until early March, he acted mostly as Second Pilot, taking the controls between take-off and landing. Night-flying commenced gradually, and Fred started acting as First Pilot more regularly. Formation and instrument flying were re-practised for the first time since Canadian training in 1941.

"Looks like you haven't lost your skills, Bright!" commented Fred L., as the Oxford remained in position in a formation of Oxfords.

"That's a relief! I was a bit worried! I really need to pass this course."

During March, Fred took the controls more frequently, gaining confidence in his instrument and night flying.

12th March 1943

For a total of nine hours, Flying Officer Wimbush instructed Fred in landing on the airstrip at night, with the guidance of two beacons. This pilot had an impressive flying record, with the potential of being promoted to a higher command. However, due to the need for experienced pilots to instruct incoming airmen, Wimbush had accepted the rank of Flying Officer-War Substantive. Pilots such as Wimbush were crucial for the inexperienced pilots being efficiently trained by the experienced.

Next came the Beam Approach, where pilots could be guided to land blindly with the aid of two audible radio beams, which, when transected, would provide the exact path for landing in almost total darkness. The altimeter would indicate the level of the aircraft above ground. Fred was First Pilot with Flying Officer Wimbush, when he was startled by Wimbush's sharp order. "Alright, Bright! Control has switched on the transmitter - listen for the tones."

Fred listened intently to the tones sounding through his headphones, correcting direction until the two tones blended to one sound.

By 23rd March, Fred had been cleared to continue on to the next step - a training course for flying Wellingtons. He was graded as an *average pilot*, and a *low average night pilot*.

Itching with anticipation, at the thought of flying an aircraft which had been fundamental in air combat, Fred waited impatiently for his next posting.

"So bloody slow!" groaned a few airmen as they looked at the blank postings board. "The war will be over before we do anything!"

The wheels of military administration maintained their reputation for turning slowly.

Those who passed their initial training were granted ten days leave so that incoming servicemen could commence their BAT (Beams Approach Training) Course. Ossington was a tiny village, with only a small pub for leisure, so Fred, in the company of other restless pilots, boarded the Midlands train to Nottingham. The reality of war hit home on their arrival. Bombed out buildings and churches lay desolately on the streets, because of a blitz in May 1941, when five hundred tons of bombs were dropped in one night. Two hundred people died from this raid, some in the air raid shelters.

Fred's former roommate commented, "I bet Robin Hood would find somewhere else to hide, if he was still around!"

The others smirked, but remained silent, taking in the traumatic signs of war.

"I suppose the shoe will be on the other foot for us soon," pondered Fred, thinking of his future bombing missions.

Wandering through the now deserted lace markets, the group separated and located several boarding houses or pubs to call home for a week.

On advice from the locals, Fred found a cave under the town to shelter in as needed. Although the air raid sirens sounded several times, no bombs fell. He could hear the "ack-ack" of anti-aircraft guns in the distance. Feeling a sense of impending danger, he clenched his jaw. Finally, the all-clear sounded and relief flooded over the locals and servicemen in the cave.

On the Saturday night before Fred received his next posting, he joined his mates and other servicemen at a local dance. American pilots were doing their best to charm the ladies, who were enjoying the attention. Miraculously, the 'Yanks' seemed to have an endless supply of silk stockings. The Commonwealth airmen had no such tempters, so had to rely on their charm. Fred eventually caught the eye of one pretty local girl, who was working in a gun factory. A drink, a dance, and good night kiss boosted Fred's spirits. It was unlikely that they would see each other again, but the night was a lovely memory to keep him warm on lonely nights.

10ʰ April 1943

No 27 Operational Training Unit, Lichfield

On arrival by train from Nottingham, Fred and the others who were seconded to this unit reported to base headquarters at Radley, in the Lichfield district, north of Birmingham. To his surprise and pleasure, Fred learned that most of the trainees were either Australians or New Zealanders. His mind eased, feeling more confident that he could head a crew more efficiently if they all came from similar locations.

Apart from delivering training on the Wellington III, Lichfield was significant, because this would be where pilots and crew would come together. Lives would depend on those choices.

Settling into his barracks, Fred looked around the room which accommodated six men, the number of air men required to fly a bomber.

At the mess, the Squadron Commander addressed the newcomers, "This is a crucial time in your flying career. This week, you will select the crew you want to be with. Just mingle, and it will sort itself out." With that gem of wisdom, the Commander left the mess.

Like new boys at boarding school, each serviceman eyed his companions, looking at insignias which identified if a man was a pilot, engineer, gunner, wireless operator, or bomb aimer. Lectures on operating a Wellington Bomber would commence the next day, so leave was granted for a more relaxed 'getting to know you' at the local pub in Lichfield village.

After a couple of pints, Fred felt as if someone was looking at him. He met the friendly eyes of another Australian who introduced himself as Bill. Trained as a bomb-aimer, Bill was from South Australia. He was two years younger than Fred. They shook hands, sensing that there was some camaraderie between them.

"Have you found your crew yet?" asked Bill.

"Nah! I've talked to a few blokes but haven't made up my mind yet. What is your opinion about this circus we've signed ourselves up for?"

"Circus is the right word, but I supposed we are all here for the same reason - get out there, bomb the bastards, get home safely, and end this bloody war!"

"Are you scared?" Bill asked candidly.

"Course I am! That's something we must live with. Fear is not going to stop me doing my job. Would you like to fly with me?"

"Yes - let's give it a go. Keep me posted about who else comes on board," Bill responded.

Grinning, Fred bought Bill another beer.

During the next few days, Fred's crew came together thanks to both Fred and Bill talking to other servicemen about each other's good characters and determination. At the end of the week, Fred's roommates had changed. His crew now shared most of his waking time. They would fly, eat, sleep, and socialise every day, with the command's theory that a necessary bond would form. The theory worked. These men would become long-term friends. In the barracks, each new crew member threw his kit on an unclaimed bed. Most of them had by now met the others but Fred suggested a formal introduction.

He started with, "My name is Fred Bright, the pilot."

"Well, you must be 'Shiny' then," quipped Pete, who was to be the navigator. The name stuck with Fred for the rest of his time with his crew.

Harry introduced himself as Wireless Operator, followed by Charlie the rear gunner.

Although most of the trainees were from the Commonwealth, Fred agreed to take on a 'pom' by the name of Walt as the Flight Engineer. He was the only man who had already flown several missions, including The Battle of Britain, for which he received a Distinguished Flying Medal. His squadron had disbanded and joined other squadrons. He had spent twelve easier months instructing new wireless operators before being posted back to Operations. His experience would help these 'sprogs' through what would be a challenging and often frightening series of missions.

"Well lads," Walt announced in a confident, calm voice, "If I can survive, so can you. It's all about working together and enjoying a beer at the pub as often as we can!"

Walt's easy going, assured demeanour gave the rest of the crew a sense of assurance that it was possible to survive, despite the daily news of fatalities.

Before the crew flew under Fred's lead, Flt Lt Frank Ormonroyd of the R.A.F was appointed to instruct Fred on the basic operation of a Wellington III. Frank had already flown several missions before taking a break as an instructor. In another eighteen months, he would lose his life in a Lancaster with the anti-aircraft 576 Squadron.

During the years after the war, Fred would ponder on all the air personnel who had helped educate him, before going to their deaths. Surviving a mission seemed to be as much a matter of good luck, as skill.

Twice a day over four days, Fred practised flying the Wellington with Frank.

"Ever been in a Wellington before, Bright?"

"Just once, as a passenger from the south of England to Egypt. I sat in as a second navigator, for the practice. She's a bit of a rough girl, this aircraft."

"Treat her like a woman. Be polite and gentle but take charge when you need to."

By the time Fred had practised corkscrews and flapless landings, he understood Frank's advice. The extreme form of flying corkscrews was essential for avoiding enemy fighters when on a bombing raid.

Frank spoke through his headphones, "I can see that you enjoyed the corkscrews, Bright! Now you will assume that you have a flap failure."

By turning downwind into the final run to the airstrip, Fred was able to reduce speed, and perform a longer landing, nose high. To his relief, the Wellington settled itself on the runway, threatening to shoot into the next paddock. With gentle braking and further reduction of power, it eventually came to a standstill almost at the barbed wire.

"Good!" commented Frank. "Now you can start taking the crew up with you. You don't need me anymore."

15th April 1943

"Right lads! Let's go! Assume your posts! We are going to fly as if this is the real thing!"

Fred did his best to sound authoritative. The crew was starting to bond, with a few small niggles related to personalities coming to the surface. Bill was worried about his soldier cousin, listed as missing. Young Charlie was lovesick for his sweetheart at home and Walt's British way of dealing with daily life irritated the more laconic ways of the Aussies. Six days of familiarising themselves with the Wellington and accepting that the First Pilot was in charge saw a more united group evolving.

The second day of crew training tested Fred's ability to deal with an emergency when the right rear tyre burst on take-off in Wellington 1905.

"Well lads! Looks like we have a small problem with a tyre!" announced Fred to the crew. He radioed the control tower. "C for Charlie- we have a burst portside tyre. I will do a go around, and attempt to land."

Summoned to the control tower, Frank, Fred's instructor spoke over the radio, "Bright, you should do a belly landing upwind. Keep the gear up. You have no bombs loaded. Fly around until your fuel is almost spent. Radio again when you are ready to come in. We will have the fire trucks and ambulance on stand-by."

"Shit!" exclaimed Charlie, sounding panicked.

"Shut up!" barked Fred. "We'll be O.K. if you stay calm. If you want to bail out, do it now."

Several muttered responses of "O.K., Skipper - we'll stick with you. If our number is up, it's up."

Fred smiled, secretly pleased that the crew was entrusting him with their safety.

Tyre bursts were not uncommon, but extremely dangerous, especially if the aircraft was loaded with bombs. Fortunately, 1905 was on a training run and not a bombing raid. An hour later, Fred radioed the control tower and announced that his fuel level was now low. On their command, Fred flew a circuit, planning to land into the wind, to slow the aircraft down as it landed.

The earth was rising up underneath the aircraft, pilot and crew.

"O. K.! Heads down! Brace! Here we go!"

Resisting the temptation to push on the brake pedals, Fred

gripped the joystick, attempted to keep the Wellington on a straight line as fuselage contacted with runway.

With a huge thump, followed by scraping and sparks flying, the aircraft finally came to a halt, landing first on its belly, then tipping slowly forward on the cockpit.

Sirens announced the arrival of fire trucks and ambulances. The men wasted no time climbing out of the bomb bay door and running clear, while the firemen doused the fuselage with water. Miraculously, no one was severely injured, apart from a few scratches and bruises incurred by hitting a waist gunner dome, navigator, or wireless desk. Sadly, 1905 had reached the end of her career due to extensive underbelly damage. After too many prangs, her parts would be used for repairs to other aircraft.

"Well done!" Frank and the surrounding admirers declared, slapping Fred and his crew on their backs.

If bonding wasn't complete before this flight, it was now. After debriefing, the group went to the mess, for a rum, to settle nerves.

Casualties and deaths during training were unfortunately too common, with seven hundred trainee pilots and crew losing their lives, even before going on active duty.

One major hazard was gas-filled barrage balloons, tethered from the ground at height sufficient to impede incoming enemy aircraft diving downwards for a bombing raid. Although this tactic was successful in deterring incoming bombing raids, it was also the cause of several fatalities resulting from aircraft flown by inexperienced allied pilots colliding with a balloon and exploding. Fred and crew, so far, were much luckier than some.

The following six weeks saw Fred and his crew form a cohesive group as they practised evasive manoeuvres, flying with one engine and emergency landings. For a second time, they went through the procedure for tyre-burst. This time, thankfully, it was practice only.

27th May 1943

Looking casually at the daily postings on the wall outside the mess, Fred saw his name, and those of his crew, stand out on the list:

Bright:
Nickel raid 2300 hours. (Operation Nickel was a code name for dropping propaganda leaflets over Germany, instead of bombs).
Briefing at 1900 hours.
Supper at 2000 hours.
Dispersal at 2200 hours.

"Crikey!" exclaimed Charlie, "we're really going on a mission."

"Of sorts," replied Fred. "We're dropping leaflets, not bombs."

At briefing, Fred knew that his Wellington 252 would be the second of three aircraft on orders to do a nickel drop, through the bomb bay doors. They would be ten minutes ahead of the Wellingtons carrying bombs.

"Get in, drop the nickel, and get out!"

At 2200 hours, Fred and crew rode the transport bus to the dispersal point and prepared for take-off. They felt exposed and naked, carrying propaganda leaflets instead of live bombs. The specially designed Monroe Bomb would break open when dispersed from the aircraft, releasing propaganda pamphlets warning citizens in enemy territory. They knew now that they would be flying over Lille, France, about three hours' flying time away, depending on the weather.

For once, all men were quiet, as they climbed through the belly of the Wellington. Each airman was mentally rehearsing his individual routine before and during the flight. On a clear night, the aircraft rolled forward, gained speed and took off to the north before circling and heading south.

As the three Wellingtons carrying the leaflets crossed the English Channel, Fred felt a sense of safety knowing that the Mosquitos were flying close behind them, and that the bomb-laden Wellingtons would also be close by.

At 0200 hours, Pete, the navigator, broke the radio silence. "Ten minutes before we see the target, Skipper."

"Bomb aimer, get into position."

Bill eased himself into the harness above the bomb bay door and armed the two bomb shells carrying the leaflets.

"Wish they were the real thing," he muttered, as flak sprayed around the aircraft.

"Bombs – no – leaflets gone!" Bill announced as soon as Pete confirmed that Fred's Wellington was in position.

A sea of leaflets fluttered downwards, warning the citizens controlled by Hitler and Mussolini's forces, that they were about to be bombed. Thankfully, by the time enemy fighters became aware of the leaflet dropping Wellingtons being present, Fred and crew had turned for home, at a lower altitude to allow the arrival of the other bombers from Lichfield. Not able to see them, the crew heard the rumble of Wellingtons heading over France.

At 0500 hours, Fred touched down at Lichfield, incident free. Their first mission was complete - a good trial run.

Two days later, the crew took off on a practice bombing run over the English Channel during daylight hours. Flying straight and steady over a target was a difficult and dangerous task, so practising in relative safety gave the crew more confidence for future missions.

After six days leave, Fred, his crew, and other crews, were being posted for the next step - training on Halifaxes. The Halifax would prove to be a fundamental part of the airmen's lives for the next eighteen months. The smell of combat was getting stronger.

Eleven

Hello Halifax!

8th June 1943

<u>1659 OTU (Operational Training Unit)</u>

Heading north to Topcliffe, about a three-hour train ride from Lichfield, Fred and his crew settled themselves into the barracks for a relatively short stay. They were feeling more refreshed after taking leave in the walled city of York, a popular location for servicemen to have a meal or a few drinks. One much frequented venue was Betty's Café, which opened in 1936. When World War II broke out, Betty, the original owner, installed a mirror in the café. On this mirror, thousands of airmen would inscribe their names with a diamond pen. On the day before reporting back to Topcliffe, Fred and the crew had finished lunch and a few ales at The Olde White Swan, and were wandering through St Helen's Square, when Walt stopped suddenly and pointed to a teahouse on the edge of the square.

"Oy! That's Betty's!" yelled Walt.

"Betty's what!" retorted Pete.

"Betty's Café! It's famous! Betty loves the fly boys! Let's have a cuppa there and I'll show you how to mark your name forever at Betty's Café!"

After finding a spare table and ordering tea and biscuits, Fred was intrigued by a group of men scrambling around what looked like a mirror.

Voices were demanding, "Hurry up!" "It's my turn!" "We have to report back in an hour!"

When the scrambling and complaining ceased after the departure of the noisy airmen, Fred stood up and wandered over to the mirror. On that mirror, was a collection of signatures, in all styles, and levels of neatness. Walt followed Fred to the mirror, diamond pen in hand.

"Ere you go, mate!" he said, offering Fred the pen. On the bottom right hand corner Fred inscribed, "F.H. Bright."

That signature remains on the mirror at Betty's Café, York, today.

10th June 1943

Fred, his crew, and ten other crews were introduced to the Handley Paige Halifax, which would be the biggest, most complex aircraft flown by Fred, to date. With four engines, the aircraft could carry six bombs at a time, as well as six machine guns. The cockpit itself presented a new challenge, due to the many gauges monitoring the four engines.

After two days of lectures, Fred and his crew started a familiarisation flight with Flight Lieutenant Harry Malkin, DFC. Harry Malkin had earlier earned his medal by successfully bringing home a badly damaged aircraft with one lost engine and non-operational rudders from a raid over Europe. Somehow, he succeeded in limping his aircraft home to a British Airfield. Harry did not like talking about his night of terror and courage, but all at Topcliffe treated him with respect and admiration.

Over the following ten days, Fred, his crew and the rest of the Halifax trainees were occupied with intense training in preparation for their final weeks before being assigned to their designated squadrons. As well as their previous exercises, they were exposed to other likely hazards of things going wrong at the worst times. Landing with three out of four engines, beam landings, and air-to-sea firing were repeatedly carried out, so that these procedures would be carried out more easily under combat conditions.

Having already heard stories of those who had taken such huge risks and survived, and those who lost their lives, all at Topcliffe were determined to enjoy their lives as much as they could before yet another embarkation. Only too aware of the fact that they would soon be in extreme danger, most of the Halifax pilots were determined to have some fun, and hopefully romance, while they could. London was beckoning! At the end of their Halifax training, Fred and his 'boys' went on a week's leave to London. They found cheap accommodation at the Royal Temperance Hotel in Carlisle, at £1-1/ per night.

Eager to find some feminine company, they had made a booking for afternoon tea at the flat of Lady Frances Ryder C.B.E., and Miss Macdonald Of The Isles, C.B.E, who had established a Hostess service for Servicemen of the Empire, and overseas students.

Lady Frances Ryder was already widely known in Australia, after an article in an iconic Aussie magazine:

> The Australian Women's Weekly, 13th April, 1940.
> *Visiting London?*
> *Go along any afternoon you are in London to 21b Cadogan Garden (Sloane Square), and ask for Lady Frances Ryder's rooms, and if you hail from any part of the Empire you are bound to meet somebody that you know there.*
>
> *Sitting around a table you might see an Australian Air Force boy talking to a Canadian soldier; a South African art student comparing notes with a New Zealand university graduate; Lady Frances herself encouraging some shy newcomer to make herself at home - or Miss Macdonald of the Isles carrying on half a dozen conversations at once, while pouring out the tea.*

These homesick young men could take their minds off what was proving to be a terrible and bloody war, at the home of Lady Ryder and Miss McDonald. Young fresh-faced people enjoyed the company of their peers in an atmosphere of grace and good manners. This experience restored a sense of well-being and optimism. Of course, the boys were looking for more than just fancy teacups and good manners. Fred had heard of a basement nightclub called 'Chez-Mois', which played great jazz, had plenty of booze, and hopefully some pretty, and friendly, ladies. What liaisons occurred were kept discreetly between friends. By the time Fred departed from the U.K. he had acquired the phone numbers of quite a few London ladies. His photograph album displays images of mysterious, unnamed, beautiful women. We can only speculate!

8th -16th July 1943
301 F.T.U (Ferry Training Unit), Lyneham

During this last time in Britain, the crews prepared for ferrying Halifaxes to the Middle East for the start of Bomber Command in that region. Now experienced in keeping a Halifax in the air, Fred and the crew tested B416 for its airworthiness, with weight testing and fuel consumption. Fred managed to squeeze in twenty minutes of night flying. They were ready.

On July 17th, B416, with Shiny Bright and crew on board, commenced the first leg of its journey to Cairo. First stopover was Hurst, in Reading, where they stayed for five days, while waiting for other Halifax crews to join them, and to refuel. Hurst was a quiet village, with none of the excitement of London. Waiting crews asked for a forty-eight-hour pass, and caught the train back to London, the place where beer and kisses were promised.

"May as well while we can, Shiny," remarked Pete. "If I have to look at your ugly mugs for the next couple of years, I need to cheer myself up!"

By now, the crew had melded as one, each man knowing that he could be called on at any moment to ensure that their aircraft stayed on target and especially stayed safe. In such a confined space in the bomber, tolerance and humour were essential, as one man would need the courtesy of others to allow him to move into his little corner of the aircraft.

22nd July 1943

In comparison to the thirty-minute flight from Lyneham to Hurd, B416, in the company of ten other Halifaxs flew a much longer eight hours to Ras-el-mar, Algeria. Because of the war-time conditions, all aircraft were ordered to fly at night along the coast of Spain, instead of over the mainland. Other Halifaxes were following at staggered distances behind, with the protection of three fighter aircraft.

Ras-el-mar was the host of 167 Maintenance Unit, where aircraft would either stop-over or have repairs completed, before returning to missions. Enjoying the temperate weather, most of the servicemen on stayover took advantage of their two spare days. After they ensured that all maintenance was underway, pilots and crew were grateful to 'stretch their legs', do a bit of sight-seeing, and explore the small town. As the population was mostly Muslim, alcohol was not freely available. The servicemen were encouraged to have coffee only, as 'guests' of Algeria. However, Fred did manage to find a bottle of beer at a restaurant. The owner was happy to bag it, after the generous 'tip' given to him after ten servicemen enjoyed his hospitality.

24th July 1943

Halifaxes and crews flew six hours east to Castel-Benito near Tripoli, in Tunisia. This airbase was built in the 1930s by the Royal Italian Air Force for training parachutists. It remained an enemy airbase until January 1943. Prior to this time, the airstrip suffered heavy bombardment from the allies, until taken over by the British Army on 23rd January 1943, and renamed R.A.F Castel-Benito. Only six months since Allied occupation, buildings were demolished and the airstrip was being repaired to flying standard, with further improvements planned. The base was now being used by squadrons for stayovers during the African Campaign. With only one night here, leave was not granted to visit Tripoli.

25th July 1943

"Getting there, boys!" shouted Fred over the roar of the Halifax's engines.

On this day they would fly six and a half hours to Cairo West Airport and refuel on an overnight stay, before the thirty-minute flight to Fayid, north of the Suez Canal.

"Things could get interesting soon!" replied Pete.

After two years of wondering and waiting, these men at last had a purpose and felt confident in their abilities as individuals, and as a crew.

With some persuasion of the base commander at Cairo West, the airmen were granted five hours leave into Cairo.

"Come back sober, and come back clean!" roared the commander, as a group of twenty men took off in a truck through the main gates of the base.

"May as well make the most of it," thought the commander, "Who knows when or if they will get another chance."

462 Squadron was waiting.

Halifax aircraft with 462 Squadron.

Bright's Boys under their Hallifax. Fred second from the right

Bright's Boys having a beer,

Twelve

The Realities of War

<u>17th September 1943</u>

"Well! This is more like it!" remarked Fred to his mates, after their second beer. "Makes you realise what we have missed. We have been living in a desert hell."

Fred, with Flight Lieutenant Cook as Second Pilot, had ferried Halifax DT 397 to 161 Maintenance Unit, Fayid, for the installation of a newly delivered Merlin engine, designed to cope with the harsh desert conditions. Fred and crew were delighted with the prospect of flying more reliable aircraft. As a bonus while waiting to ferry an aircraft back to base, they also could enjoy a few days of full showers, and food other than bully beef.

22ⁿᵈ September 1943

"Meet JBB419, boys! She's all yours. Fuelled up and ready to go!" declared the maintenance supervisor. This 'Halibag' would be Fred's designated aircraft for several future missions. JBB419 was dubbed J for Joyce.

"That's the name of one of my girlfriends!" boasted Fred.

"Hope she's not a tart then!" quipped Bill.

Fred did not answer.

23ʳᵈ September 1943

"O.K. boys! We're back on ops tonight! Let's see how J for Joyce behaves for us! This mission is fair dinkum - no flares. Tonight we are delivering the goods."

The usual quiet before a mission had settled over the squadron, with airmen writing letters home, 'just in case.'

> *Dear Mother and Dad*
> *I am in this hot, dusty place, still in one piece. We had a short break last week and enjoyed a few cold beers while we could. We have a mix of Aussies, Poms and Kiwis here. Each group thinks that it is superior, and that the squadron is either R.A.F, or R.A.A.F, depending on what side you are barracking*

for. It can make for some interesting conversations! When we aren't flying, we have started a cricket team for entertainment, late in the afternoons, when it is cooler. With the Poms playing against the Aussies, I think we could start our own war! Some wag managed to bring a piano out here, in the back of beyond. If you knew where we are, you would be amazed that a piano could survive. Still, a bit of a sing-song is a good thing, when we are waiting for something to happen. I hope you are both well. Thanks for your letter. It took two months to arrive, but news from home is always good, no matter how old it is.

I will write again soon.
With love,
Your son Fred.

By the time Fred had finished his letter and left it on the end of his stretcher, he was called to the pre-mission briefing, have supper, and go to the disbursement area. All the aircraft were encircled around the tents, to lessen the chance of being blown up by enemy aircraft.

On this night, the target was the Kalathos Aerodrome, on the island of Rhodes.

At twenty-five minutes after midnight, J for Joyce, in formation with five other aircraft, followed the flare plane towards Rhodes.

"Visibility is poor, Skipper," commented Pete.

"Roger! I see that!" came the slightly sarcastic reply.

As they approached Kalathos Aerodrome at a level of 12,000 feet, Bill spoke over the intercom.

"Hard to see the runway, Skip! Too much anti-aircraft flak."

"When you think you are close enough, bombardier, disperse half your load."

A few seconds later, the bomb doors were opened and then the first half of eleven, 800 pound bombs were dropped.

"I think we were a bit off target, Skip."

"Let the rest go when visibility improves – should be in a minute."

Two minutes later, at four thirty-four am the remainder of the bombs were released, successfully striking the far end of the runway.

"Let's head for home, boys. Not a bad effort for a first bombing run. We'll do better next time."

Each of the six aircraft on the mission had similar issues, with worse results. One aircraft had engine failure, and although returning to base safely, burst into flames. The second aircraft was hit by flak, lost height and had to jettison its bomb-load, and return to base. Secretly, Bright's Boys wondered when/if it would be their turn to face such perils.

They attacked Kalathos the following night, with more success.

During the next three nights, the squadron bombed several other Greek airfields: Kalmaki, Larissa, and Argo. Fortunately, J for Joyce returned each morning, unscathed.

At debriefing after their fifth mission, Fred commented, "I think it is only going to get tougher from now on. We can't be thinking about 'buying it'. When your number is up, it's up."

To their relief, several crews were stood down for a few days. This gave them a chance to get away from the dusty confines of their tents and relax at Appolonia, on the north Benghazi coast.

"Clean sheets! No bed bugs! Eggs! Bread! Fruit! I'm in Heaven!" cried Bill.

The relief of doing something relatively normal for a day or two refreshed body and soul. A tidal swimming pool provided much needed relief from the relentless desert heat.

Climbing on board the transport truck for return to duty, Bright's Boys were ready to take J for Joyce back to work.

The next two months would test the skills and courage of all members of 462, as the enemy desperately defended the Greek Islands.

2nd October 1943

J for Joyce's bombing run over Kalathos Aerodrome was successful, despite cloud obscuring the target. Once again, she brought Bright's Boys home unscathed.

The following night, six Hallifaxes flew over the Maritsa Aerodrome, Rhodes, with a load of eleven 500lb bombs. Although the bombing run was successful, five anti-aircraft guns had J for Joyce in their sights, although only one gun found its target. Fortunately, no major damage was done, but the nerves of Fred's crew were shaken. Fred concentrated on bringing his men 'home' safely, and attending debriefing, before he allowed himself to feel anything. Even then, he put on a facade of indifference by whistling tunelessly, much to the annoyance of anyone within earshot.

"Bright! Can it!" sounded a grumpy voice in the mess. With that, Fred retreated to his tent, exhausted from seven hours of flying, and dodging bullets.

The next three missions over the Aerodrome at Rhodes were successful, with no flak holing the aircraft, although the spray of bullets was seen by Bill. On his warning, Fred called out, "Hang on" and cork-screwed out of their path.

"Those bastards aren't getting me a second time!" he vowed to himself.

A false sense of security had returned, but it was short lived. A second run over the Maritza Aerodrome resulted in more flak hitting J for Joyce. Once again, Bright's Boys returned to base safely. While the risk of not returning was present in each man's mind, the missions demanded their full attention. Apart from a few jokes about 'kicking the bucket', any conversation that touched on the fact that some would not come home was considered bad luck, and not spoken about. Even after the war, Fred rarely talked about his experiences, or his fears.

Although Bright's Boys were stood down on this particular night, the fury of the Luftwaffe over Maritza was all too evident. R.A.F pilot Sergeant Mike Hall, and crew on a flare run, were shot down by ground fire and plummeted into the sea. All on board were lost. Fred noted in his log book; *X-XRAY- Sergeant Hall Missing.*

A second Halifax also went down. R.A.F Sergeant Marsh and crew, on their first mission with 462, while on a circuit after take-off, started losing height, finally crashing one and a half miles away, with a full load of bombs. The explosions and fire incinerated all on board. They would not be going home. The details were not immediately known when, once again, Fred noted in his log book; *A-Apple - Sgt. Marsh Went in?* He later wrote; *KIA*

(Killed in Action) Oct 14th. Missions were scrubbed for two days to allow the now close-knit airmen time to reflect and mourn their lost mates.

During the following nine days, J for Joyce flew more missions over the aerodromes of the Greek Islands. Although the anti-aircraft fire, and Luftwaffe fighters remained fierce, Fred managed to keep out of their way.

26th October 1943

Fred had a welcome break from flying missions by ferrying another aircraft to Cairo West, with R.A.F Wing Commander Bill Russell, who had been Commanding Officer since August 1943. He was highly regarded by the squadron, as he had taken steps to improve the living conditions of the longest serving airmen, as they were suffering from malnutrition and scabby sores. He arranged for groups of men to have a break from the poor conditions at a coastal town, with better food available. The men would return refreshed, with morale restored, ready for more missions. Under his command, 462 Squadron Middle East gained the reputation of being a dedicated unit. Bill Russell enjoyed the opportunity of flying again, so Fred happily acted as Second Pilot. Embarrassingly, the aircraft had engine failure, and returned to base. The flight resumed the following day, this time successfully. Although Fred would not have known why he had been ordered to accompany W.C. Russell to Cairo, he would find out when he returned to base.

27th October 1943

Leaving Wing Commander Russel in Cairo, Fred flew from Cairo to Fayid, with Sergeant Heard, a pilot who also had the same experiences of the Greek aerodromes, as Fred's. Each man felt a sense of relief when talking, at least at a superficial level, about the dangers they were facing. At Fayid, each pilot would meet with his crew, for return to base.

28th October 1943

Fred, happy to be reunited with his crew, prepared for a mission from Fayid to Cinque Terre, Italy, on JBB419. When the inner port engine failed on take-off, they returned to base. Once again a frustrated pilot and crew instead ferried W7849 back to base.

On 29th October 1943, JBB 419 was declared airworthy, so Fred and crew prepared for a mission over Heraklion. Once again, the inner port engine failed. Fortunately, Fred managed to bring himself and Bright's Boys back to base, without incident.

"Is there anything you can do with this engine, George?" Fred asked plaintively of a ground crew mechanic. He just wanted to get the war over and done with now, preferably in one piece.

"It's sand in the spark plugs - I'll see what I can do."

George and his crew spent the following day dismantling and reassembling the recalcitrant engine in readiness for its next mission.

Things just became steadily worse. On their return to base the following day, Fred and his crew quickly became aware of the reason for the Wing Commander's visit to Cairo - 462 Squadron was on the move.

'But we just got here!" Pete complained.

"Didn't know you had become so fond of the place, Pete!" commented Fred.

Italy had surrendered on September 8th, 1943, so 462 Squadron was ordered to move to Malta, to assist other allied squadrons in weakening Hitler's defences. Everyone was happy to get out of the desert and see some greenery around them again.

The orders were to uplift all tents and facilities, while continuing to fly bombing missions at night. All tents had been dismantled and packed, ready for returning to 161 Maintenance Unit, but nobody was moving.

"What's going on?" asked Fred. "I thought we would be preparing to fly to Malta by now."

One airman, who had been finishing a cigarette under the wing of a Halifax, walked out into the sun and ground the stub in the sand, grinding it angrily under his foot.

"Malta has been scrubbed! Now we have orders to go five miles down the bloody road! We flew missions last night, packed this morning, and now those morons have changed their bloody minds!"

This last order was the third one in three days. Nobody felt like moving, but the desert heat forced the exhausted squadron members to continue with the decamping and move, as ordered, five miles down the road, to set up camp again.

Failing engines, diplomatic disruption and the difficult living conditions were taking their toll on all the squadron. In true style, Fred merely jutted his chin out, determined to get himself and his boys through, no matter what. From the new base 'five miles down the road,' missions recommenced as soon as camp was set up yet again.

2nd November 1943

JBB419 obviously had had enough of flying as well - the inner port engine started misfiring, and was declared unsound, twenty miles before the night target over Heraklion.

Flying over the ocean, Fred ordered, "Bomb doors open! Jettison all bombs!"

The lighter load, and lower chance of explosion ensured that Bright's Boys, even if very disgruntled, should land safely.

J for Joyce landed at Benina, Libya, as a safety measure. The next morning after take-off to base, the engine failed completely. Fred brought Joyce down at base on three engines.

Just as the squadron was valiantly re-settling and adjusting to the loss of mates, disaster struck again in the worst way.

20th November 1943

Halifax BBB431 H for Harry, exploded on the base while being bombed up. A faulty bomb fuse was later blamed as the cause. It was impossible to tell who had been affected by the massive explosion. A roll call was immediately ordered - eighteen men were missing, killed instantly. In one moment, air crew and ground crew were joking around the doomed Halifax, and then…life was over for them. This was the second such tragedy in seven months. Morale was non-existent by now. Fred joined his mates in a grim search through the debris to find the remains of these men for burial the following day.

Such was the despair of 462, Headquarters ceased all missions until early December, so that routines could be more relaxed, with leave granted to visit Benghazi for rest and recreation.

The base itself was tidied and re-organised, so that life was more bearable. Arrangements were made for sports teams to be organised, to allow the men to let off a little steam. Rugby and soccer were very popular, as was table tennis, darts, and draught competitions. To the end of his life, Fred was an unbeatable table tennis player.

Entertainment troupes visited, as well as the men creating their own dress up plays and concert. It was a sweet relief to have such distractions, although many of the men were having nightmares about the recent tragedies.

It was at this time that Fred seemed to be having too good of a time. One evening, after a few glasses of the local beer, he was walking unsteadily to the latrine, not seeing a kerosene tin, which had the top cut out, to form a bucket. Fred fell over it, and crashed to the ground, swearing loudly, waking a few sleepy airmen.

"What the hell is going on out there!" Heads poked out of tents to see Fred on the ground, trying to stem the flow of blood from a long, deep gash below his right knee.

"Well, that's buggered my flying! I will be right when we go back on ops."

The base medico differed. "You will need about ten stitches, and bed rest, Bright, before you get back in a kite again!"

With his leg sutured and bandaged, Fred was transported by the volunteer ambulance to the British Medical Hospital, Benghazi, for the prescribed bed rest and penicillin, if it was available. The one hundred bed hospital was staffed only by twelve nursing sisters, so Fred had to exercise patience during the first week, until he could move around on crutches.

His crew visited him a few days later.

"Well Shiny, you are going to miss out on all the Christmas fun the CO is planning for us."

"Bastards!" came Fred's reply.

Christmas and New Year brought on a bout of homesickness for Fred - his family, but especially his other family, his crew. "What are they doing now? They had not better be enjoying themselves too much without me."

As all the patients and nurses were also away from comrades and families, Fred decided to stop feeling sorry for himself and attempt to at least flirt with a couple of the nurses, or chat to the chap in the bed next door. He had burned his eyes in an explosion, and was temporarily blinded with bandages. Fred filled his time reading any old newspapers or trashy paperbacks to him.

January 1944

Fred's return to 462 Squadron saw major changes happening around him. The desert campaign was almost over, and sights were now on using Italy as a squadron base. While some airmen continued with missions for a few weeks from Libya, others were deployed back to Britain for secret training. The remainder packed and embarked for Naples.

Between early January and late February, Bright's Boys completed four missions over Greece and two air tests. These flights marked the end of this intense time of strong friendship, which gave the crew the courage to fly with Fred without question. As these remarkable men prepared to leave for their new postings, Fred would go on alone. On the last night, in Cairo, they spent many hours drinking, singing dirty ditties, and reminiscing. The next morning, it was time to part company. No man would admit to the tears in his eyes, or the lump in his throat. With much back-slapping, and then, finally, a salute to Fred, their Skipper, it was all over. None of them would ever be the same again. To rely on mates during such dangerous and frightening times, the bond they had forged would never break, even though not all of them would meet again.

With their thirty missions completed, the crew would be posted to less hazardous roles. Bill, Harry, and Second Pilot Martin stayed on in Egypt until 1945 to assist with the still necessary defence and security of the Middle East. Likewise, Pete (the Pom), was posted to Dumfries, Scotland to the Advance

Flying Unit, where he instructed gunnery. He later re-met with Fred, in Toowoomba, Australia. Walt left the crew in late October, presumably back to Great Britain. (The author has been unsuccessful to date, in tracing his life after 462 Squadron Middle East.)

Fred, with a group of forty other airmen, embarked on a troop ship for Italy. For the first time since he enlisted, he felt lost.

Thirteen

Over the Mediterranean

9th February 1944

"O.K. boys! One more mission for the Allies!" declared Fred, as the crew prepared for what would be their last mission from the Middle East. Halifax HBB420, with four other aircraft, had been ordered to provide diversionary bombing over Candida Harbour, Crete.

"We're just cannon fodder on this run, but at least the other boys should be able to do some real damage!"

Four hours later, all aircraft returned safely.

Their adrenalin rush faded into a bored lethargy as Fred remained at El-Adem, Tunisia, with other aircrews waiting for the construction of an airstrip at Foggia, near Croton, Italy. Although missions continued, Fred was ordered to temporarily stand down. He had completed twenty-nine missions, necessitating a break from the mental stress of aerial bombings. Not taking off at night with his crew for enemy territories felt strange to Fred.

After Mussolini's surrender in September 1943, the British Air Ministry reassembled its overseas squadrons. After an intense two years in the Middle East, 462 Squadron was listed for relocation to Driffield, Yorkshire. Higher command planned to utilise the Middle East Halifaxes for heavy bombing over Germany, out of Yorkshire. The change of 462 Squadron would occur from Italy, early 1944, when the crewmen remaining behind would be recognised as 612 Squadron. The Australian flight crews remained in the Middle East, much to their disappointment. From a total of two hundred ground crew, groups of forty would be gradually posted back to either Britain, or returned to Australia on medical or compassionate grounds. Already seasoned in warfare, most of them were ready and willing to continue with bombing raids. Apart from the regret of not staying on in 462, there was also the sadness of the parting of British and Australian airmen. Fred's navigator Pete was seconded to 203 Squadron, which had commenced patrols over the Bay of Bengal.

"...and although at the start, there was a lot of ill-feeling, this has been overcome, and at the present time, there is a spirit of comradeship and mutual respect in the squadron which will be very hard to replace. Everyone on the squadron will miss them very much." (Unit historian)

Between September 1942 and March 1944, 462 Squadron flew 830 missions, driving back Rommel and disrupting the Nazi advances. Sixty-three men and twenty-six aircraft failed to return. Fred was no longer a young adventurer - he had grown to be a man who had witnessed the best and the worst of human behaviour.

> Dear Mother and Dad
> Thanks for your three letters. They all arrived at once, so I read them in order. Fancy Heather nearly finishing school! I won't know her when I see her again.
> Hope Dad's gout improves.
> Sorry I have not written to you recently. I have been kept rather busy, but still in one piece, apart from a run-in with a kerosene tin! Don't worry, my leg has completely healed, and I am flying again. Stupid bloody thing to do! My squadron is being broken up - we are all very sad about that, as we have formed a great group. The C.O. asked me if I would like to be Squadron Leader a while back before, but I said 'no' as I wanted to stay with my crew. As it turns out most have already been posted elsewhere. My navigator Pete will be coming with me. Sadly, the tail gunner has been granted compassionate leave, to help his sick father on the family farm. I miss them a lot.

They were a great bunch and good friends. I hope we catch up again one day.

I am being posted again, to a nicer location. I don't expect to be in too much danger, as my duties will be more routine from now on - or so I am told! It is going to be quite chilly in my new postings. That will be a shock after the dry heat of the last year or so.
When I get settled, I will write again.
Love from
Your son Fred.

Although Fred did not fly any more than his previous missions from El Aden, he was kept busy helping aircrews to "bomb up," and prepare for flight. Every man was mindful of the earlier tragedies of bombs exploding before take-off, so nerves were ragged.

To add to their misery, the worst dust storm in memory raged through the base.

"...Worst of all, the mess tent collapsed, and tins of food had to be handed out so that people could eat in their own tents. (Unit historian*)*

Just as the small group of airmen was setting the base in order again, in came some crews, which had been previously sent to Britain, to train on new radar targeting equipment; the Gee navigation system and a Mark IV Bombsight, which allowed more accurate bombing from a lower height of 3000 feet. A mechanical computer would assist with bomb weight, velocity, and wind speed. The incoming crew then trained those left behind in readiness for extending missions from reclaimed Italy territory further eastward into Nazi occupied Europe.

"Will this ever end!" groaned Fred to James, an English pilot.

"Chin up, old chap!"

Fred walked away, muttering, "Oh shuddup!"

James, fortunately, did not hear that comment.

Finally, orders came through, to relocate to Foggia Airfield, Creton. The last Mission for 462 Squadron from the Middle East was flown on the night of 23rd-24th February, 1944.

At 0640 hours on 28th February, the remnants of 462 Squadron Middle East took off for Foggia Airbase, landing on the new, short strip at 1600 hours. They were now happily reunited with their devoted ground crew, and finally out of the desert. Fred took a new crew, and one passenger with him, in Halifax II HBB420. (H.B. means Heavy Bomber) The bomb damage at Foggia was so bad that tents were again the necessary accommodation.

"At least this time we don't have bloody sand!" mused Fred.

7th March 1944

This would be the last day when Fred would fly a Halifax, an aircraft which, despite its challenges with engine failures, he had grown to love. He was pleasantly surprised to see the former 462 Squadron Leader, Bill Craig, who would accompany him on an air test of HBB420.

After the thirty-minute flight, Bill said, "Well, Bright, this flight concludes your Operational Tour with 462 Squadron. You have flown well with your missions. I am going to recommend you for a commission. It's time for you to start instructing the new sprogs coming in. You have excellent leadership skills. I'll get your log book written up. Welcome to the 205 Communications Group of 614 Squadron."

Fred had a total of 628 hours, ten minutes of flying experience in the three years since enlistment, 486 of those hours were operational.

A week later, Fred, and several other pilots recommended for promotion, or a Commission, were lined up outside the Wing Commander's office, waiting for an interview.

"Warrant Officer Bright!" barked the sergeant.

In his new dress uniform, Fred marched into the Wing Commander's office and saluted to the four senior officers seated behind a long table. He stood at attention, looking at the wall behind the table.

"Stand easy," a familiar voice ordered.

Fred relaxed a little and dared to look at his four interviewers. He was relieved to see that two of them had been Squadron Leaders at 462 Squadron Middle East.

Group Commander Kenny, whom Fred did not know very well, spoke: "Well, Bright, I want you to tell us what your ambitions are, and why you think you would make an efficient Commissioned Officer."

"Sir, I have always loved flying. I have found serving the Commonwealth through my enlistment with the British and Australian Air Forces an honour. As I have learned to fly a series of aircraft, especially the Halifax, I have also learned to be a good leader to the men who flew with me in the Middle East. We each knew that we could all depend on one another in dangerous situations. Although I have completed my operative tour, I am keen to instruct less experienced pilots in what I know."

"I see, thank you."

Bill Craig then spoke; "Bright, what are your plans after the war?"

"It is my wish to remain in the air force, sir. It has become my life.".

Each of the four officers, who had been making notes during the interview, leant in towards each other, conferring quietly. At last, Bill Craig gave a half smile and nodded. "Thank you, Warrant Officer Bright. That will be all."

The wheels of military bureaucracy turn slowly. Fred's missions were acknowledged five weeks after their completion.

In the meantime, a superior officer lodged the recommendations for Fred's commission.

26th March 1944

Wing Commander W. Russel wrote:
A keen young NCO who has almost completed his first Ops Tour. He is enthusiastic and should make a good officer. Recommended for a commission in the GD (General Duties) Branch.

Also on this date, Fred was promoted to Pilot Officer. The Commission would take much longer, to be formalised.

16th April 1944

Group Commander Kenny wrote on the same form;
A keen Warrant Officer possessing drive and command, who is

desirous of continuing in the air force after the war. Recommended for a commission.

18th April 1944

Log book inscription:
Ability as an HB pilot: Average; Comments: Has completed a good operational tour.
Signed W. Russel, Wing Commander.

19th April 1944

Group Commander D. Mcnair concurred; *Recommended*

Finally, on 29th April 1944, Flight Lieutenant Greeves confirmed Fred's flying hours as being correct.

20th May 1944

The Ministry of Defence received the recommendation for Commission.

In the meantime, while the Commission process was in progress, Fred was granted a long-awaited 'rest break' in Sorrento, at the Hotel Minerva, Italy. He and a group of other airmen either having completed several missions or about to commence them, were also desperately in need of some comfort and relaxation. The journey, however, was arduous. Starting out at the Foggia railway yards, the lucky ones on leave boarded an ancient steam train, with wooden carriages and wooden seats.

Fred fidgeted, or stood up when he could, as his desert missions had left him lean, with no padding for comfortable sitting. Although the train was operational, the station buildings were partly demolished from constant air raids. Outside Foggia, the poverty and misery of the Italians of the south were heart-wrenching. Each small farm was barren, with starving stock. The peasants themselves also suffered from malnutrition, no hygiene, and abject apathy and misery. All on the train, looking forward to some relative luxury, watched the passing scene in a gloomy silence.

The train went as far as Naples, where the airmen would spend the night, before moving on to Sorrento. The once beautiful city was broken, with demolished buildings, and families sleeping in the streets. Some were begging for food. Morale was slipping steadily as the airmen witnessed the horrors of war at ground level. It was far more confronting than viewing a target at 12,000 feet.

In contrast the next day, their sadness turned to terror and laughter, as the bus took them downhill from Naples to Sorrento on the coast. The driver did not seem to have brakes - the further down he drove, the faster the speed, around jagged curves, perilously close to the cliff edge above the coast.

The airmen banged into each other.

"These are Come Over Darling Corners!" shouted one man. "Shame I don't have a girl beside me!"

At last, the Hotel Minerva, overlooking the beautiful Bay of Naples appeared. For no cost, Fred and his holiday companions enjoyed the luxury of comfortable beds and 'real' bathrooms. The Italian staff served them as if they were paying guests. The wine flowed - life was good again, at least for a while. Fred enjoyed swimming, eating, drinking, reading, and sleeping. The local citizens were friendly, and grateful for the Allied presence in their region. Morale rose - laughter and chatter echoed throughout the hotel.

All too soon, the week was 'up', and it was time for the return journey to Foggia. The charter bus delivered the next crews on leave and collected Fred and his new friends. Just exchanging news with the exhausted newcomers made Fred realise how very badly he had needed this break. Although the Allies were still taking control of parts of Italy, they now also focused on Eastern Europe, reaching as far as Yugoslavia and Bulgaria. The aim of the personnel at Foggia was to disrupt the infrastructures of railways, oil supplies, bridges and factories which were essential for the Nazis to continue fighting.

Over several months a small number of squadrons from 205 Communications Group supported the much larger, 80 Squadrons of the United States Fifteenth Air Force. Collectively, they were known as the Mediterranean Allied Strategic Force. Enemy resistance was fierce, with high loss of aircraft and crews.

Although Fred was disappointed to be excluded from this campaign, he was also aware that his risk of injury, or worse, would increase, with the odds lessening, and battle fatigue impairing his judgement.

He waited, yet again, to hear word about his next 'real' posting. In the meantime, he retained his flying hours by taking the controls of smaller aircraft, including the four seat, single engine Fairchild, and twin-engine transport Anson, on 'milk runs' between Brendisi in the Naples region, and Termoli, close to the Foggia base. He carried pilots about to start their operations, as well as those who had completed their missions. Most of them enjoyed flying smaller aircraft as a diversion. Fred happily agreed to be Second Pilot on several occasions.

Trying to contain his envy of their recent flying action, Fred commented to Sergeant Merrick, also at the end of his operational tour, "It is hard to stop flying missions. I miss the crew and the adrenalin. At least my mother will be pleased to see me when I get home."

Sergeant Merrick nodded. "I know what you mean, Bright."

29th April 1944

Fred was posted back to the Middle East, to wait, again. He left Foggia for El Ouina, Tunisia, three hours flight south. For once, Fred found himself on a Douglas aircraft, commonly known as a DC3 or Dakota, according to the R.A.F. Originally a civilian transport plane, it could carry twelve passengers. During World War Two, it was used for military transport. Its few quirks such as leaky windows and a challenging brake system resulted in it being dubbed "The Goony Bird". Fred felt quite strange, not being at the controls of an aircraft, instead relying on the skills of the pilot, unnamed in his log book. Like Fred, other 'passengers' were also departing Italy on the Goony Bird.

On 1st May 1944, they touched down at Castel Benito, Libya, a short two-hour flight away.

Another night stayover, and the Goony Bird took off for the five-hour flight to Cairo West, the R.A.F base. Fred would stay for the next month; he was at the No 3 Base Personnel Disposal, designed for airmen who were O.T.E (Operational Tour Expired, when they had completed their mission.)

Feeling even more redundant and despondent, Fred lay on his bunk wondering, "Am I that bloody useless?"

The general 'go' was that this was where expired pilots 'just buggered around until Headquarters found something for them to do.'

Fred asked his Commanding Officer, "Now what, sir?"

Ignoring Pilot Officer Bright's abruptness, the C.O. replied, "We have plans for you, Bright. You won't be idle for long."

Dear Mother and Dad

I am back in the land of heat, sand and flies, with not much to do. I am waiting, not so patiently, for orders to come through. You will be pleased to know that I have been recommended for a commission, although it most likely takes a while to come through. I'm grounded at the moment, as other crews are using available aircraft on 'necessary tasks.' I've got a desk job at the moment, sorting and organising leave requests. When I'm not needed, I go into town with some of the lads on the base. I may as well see if I can find a few presents to send home for Christmas, while I can. This war is dragging on, but hopefully for not too much longer. Morale has been good, but everyone is tired and ready for home. All going well, I will be back in the air again soon.

Hope everyone is well at home.
Love from your son Fred

11th May 1944

Much to his relief, Fred was finally 'given something to do.'

Squadron Leader William Craig was as good as his word; Fred was about to start an instructor's course on flying Liberators. He had two weeks back in the classroom, first being re-assessed on his own acquired theoretical knowledge, and then being trained on the specific features of the Liberator heavy bomber.

Although the Liberator was a heavier bomber, it was like the Halifax in that it was a four-engine aircraft, which could fly long-range, and carry up to six tons of bombs. The cockpit was more comfortable than that of the Halifax, as the Second Pilot seat was now next to that of the First Pilot. Although the basic operation of the B-24 was very similar to the Halifax, its controls were much more sophisticated and complicated. Fred carefully studied a film entitled, "How to fly a B-24 Liberation Bomber."

"Look at the size of this manual!" he exclaimed to a classmate, "I'm looking forward to getting this baby in the air!"

25th May 1944

Fred found out very quickly, what a joy it was to fly this relatively new aircraft. He acted as second pilot over four days, familiarising himself with the feel of the B 24 and its longer pre-flight and post-flight checklists.

Flight Lieutenant Pink, who had already flown several missions on Liberators, was very helpful in training Fred with practical aspects of flying this aircraft - circuits and landings, bombing practice, and air to air firing.

Fred itched to get into the pilot's seat and fly more missions. This desire was not to be. After a few practice flight tests as First Pilot, on 1st June 1944, Fred Bright became an Instructor Pilot directing each student from the second pilot's seat.

After two more days of practising 'touch and goes' (circuits and landings) as a second pilot, Fred took the first pilot's seat, with a sense of satisfaction and excitement. Although flying was still dangerous, the fact that they were not in combat made the exercises more enjoyable. Of course, protocol was strictly obeyed, as the men whom Fred helped train, would be going on their own missions soon enough.

From early June to late October, Fred instructed over sixty days, teaching approximately forty pilots, in a number of Liberators. During the instructional period, any mechanical issues could be identified, before going on missions. Learning to fly the aircraft from the co-pilot's seat was particularly important, in case the first pilot became ill, injured, or worse. This aircraft needed two pilots to manage the controls, so learning to fly solo from the right seat was crucial. The central controls were accessible in the centre of the cockpit, enabling the co-pilot to take control, in order to land safely.

Fred briefly reunited with his own second pilot, Pete, who flew with him, just once, in a Liberator. Whether Pete continued to fly missions in Liberators, Fred did not know.

One of Fred's most renowned 'students' was Flying Officer Peter Raw, who went on to fly Liberators over northern Italy, with the 178 Squadron, Foggia, Italy. He later undertook several extremely perilous missions over Warsaw, delivering supplies to starving Polish civilians.

Subsequently taking command of 205 Communications Group, he was awarded a Distinguished Flying Cross. His military career continued in Britain and Australia until 1978, when he retired, as Air Commodore.

"Well," thought Fred after Peter Raw left to join the 178 Squadron, "At least if I can't go back to flying missions, I know that this bloke will be a bloody good leader, as long as he keeps his head down!"

During these last few months in the Middle East, the only incident reported was that of a clip missing on an upper mid-gun turret on a Liberator. Fred had come through his missions alive, and mostly unscathed, physically. The psychological effects would stay with him for the remainder of his life.

28th October 1944

Fred flew his last instructional flight with Flight Sergeant Smith - a routine Circuits and Landings Exercise.

Wing Commander Harris wrote in Fred's file:

A very reserved and quiet officer interested and capable in his work as an instructor. Still rather inexperienced, but his obedience and manner of carrying out his duties give him promise of development.

30th October 1944

Fred transferred to 21 Personnel Transit Centre at Kasafareet, the Middle East, waiting for transport 'home'.

Having a beer with a pilot also waiting, Fred commented, "Now that I am so close to going home, I don't know how I feel - happy? Sad? This has been a bloody war, but I made some great mates here. Wonder if I will ever see them again."

"Yeah, mate, I know what you mean. I feel the same," replied Fred's companion, slapping him on the back in support.

8th December 1944

Fred flew as a passenger on a Dakota from Cairo to Dar-es Salaam, a Tanzanian port on the East Coast of Africa. He boarded the SS Hindustan, a British transport ship, on 18th December, sailing down the coast of Africa to Cape Town, onwards to India, and then down the West Australian coast, across the Great Australian Bight, before arriving in Melbourne on the south-east coast, on January 8th, 1945. He was home. On the ship, stringent rationing resulted in a very modest Christmas lunch, with bully beef and beer.

The war was not yet over - Fred fervently hoped that his flying days would continue. Luckily, he wasn't grounded, just yet. He would continue flying for two more years.

Fourteen

So What Happened to Fred?

<u>1st February 1945</u>

The front door at Hume St, Toowoomba creaked open, and a familiar voice called out, "Anyone home?"

With a shriek, Heather galloped down the hallway, into her brother's arms.

"Well! Look at you! You've grown up a bit!"

"Mum! Mum! Fred's back!"

The sound of feet running from the kitchen echoed along the hallway, towards where Fred was still standing.

"Fred!" Clara sobbed, clutching her son in a way that said she could not let him go, ever again.

His father's voice behind Clara said, "Well, I see you are home again. Pleased that you're still in one piece."

Apart from the clearing of throats by father and son, a handshake, and a slap on the back, not much else was said. Fred was desperately trying to stop his chin from wobbling with emotion *(a very Fred mannerism)*.

Later, in the kitchen over a cup of tea, Fred said, "I'm home, but still in the air force, so will have to report back to Laverton in Sydney by the end of the month. I am going to be a glorified taxi driver. Still, I will be flying, so that's a good thing."

Clara looked pleadingly at her only son. "You will be going back to your old job when the war is over, won't you," she said more as statement, rather than asking a question.

"We'll see."

Fred badly wanted to stay in the R.A.A.F, or at least fly with the Civilian Air Force, but he wisely decided to not upset his already nervous mother. The war in Europe was waning in favour of the Allies, while the Pacific War was continuing to drag on. Everyone was war-weary, and still on edge. Life would never return to the pre-war 'normal'.

To Fred, however, life suddenly seemed slower, calmer, and, after catching up with family and friends, a little dull. Those at home had no idea what he had experienced, or how it had changed him. He still felt on constant alert, ready to fly another mission.

At the end of his leave, Fred felt a sense of relief at returning to what had become so familiar and dear to him in the almost three years since he enlisted.

9th March, 1945

Fred commenced duties at the Test and Ferry Pool, Laverton, Sydney. During March, he did test flights with other pilots in the Sydney area, flying Oxfords, or his new favourite aircraft, the Liberator.

Between April and October, Fred flew Liberators as far as Pt Moresby, Townsville, Charleville, Alice Springs and all points in between. He mostly had stayovers of only one night, before continuing to the next destination. He usually had the company of another pilot, who would share the pilot duties. At each stayover, grateful locals would insist on shouting (buying them a beer).

A major part of Fred's role included delivering Liberators to R.A.A.F Amberly base, outside Brisbane. Number 12 Heavy Bomber Squadron was forming, and ultimately acquired 254 Liberators.

Like Fred, Australia had also changed through the War. Darwin and Broome had been bombed, with the R.A.A.F at home vigilantly defending north Australia. In the south, enemy submarines were discovered in Sydney Harbour. During his transport flight duties, Fred had time to think about his combat experiences. He felt proud of his achievements, knowing that he had helped to protect his homeland. Yet, the damage he witnessed in Italy continued to haunt him.

As the War was not yet over, combat conditions remained. Although Australia dodged enemy invasion, or a repeat of the London blitz, people were apprehensive and determined to survive and protect what they loved. Bomb shelters lay beneath the middle of Ruthven St, Toowoomba's main street. The railway station, hospitals, schools and suburban back yards also provided protection. Blackout conditions, requiring all windows to be covered with black curtains, remained. Rations would continue until 1950 as the economy recovered, and later boomed.

8th May 1945

Victory in Europe

Fred was back in Laverton when peace in Europe was declared, All unnecessary flights were stood down so that most aircrew could quietly celebrate and relax, knowing that the war was almost at its end. As the Pacific War continued, no public revelry happened. Over many beers, Fred reminisced with those who knew what life had been like for so many courageous men and women. He looked sideways at a fellow pilot. "Well, Phil," he said, with a slight slur, "This is bloody good! We got the bastards!"

"Yes, mate, we did get the bastards! Another beer?"

14th August 1945

With the dropping of the atomic bombs on Hiroshima and Nagasaki in Japan, Allied Victory was at last declared.

2nd September 1945

Japan, which never wanted to admit defeat, formally surrendered on the foredeck of the U.S.S Missouri, in Tokyo Bay. In the brief ceremony, Allied representatives and General McArthur signed Japan's Declaration of Surrender, co-signed by Japanese delegates.

The Liberators were now utilised for peaceful defence on several R.A.A.F bases. Fred continued delivering Liberators to Amberley.

With air servicemen returning from bases scattered all over the country, Fred, and other pilots with the Test and Ferry Unit spent the next two months transporting them from remote locations in smaller aircraft such as the twin-engine Anson. The Anson was used during the war for maritime reconnaissance. One of the first models with retractable landing gear, it was not suitable for combat missions, but continued post-war to facilitate civil aviation. This smaller aircraft was proving to be a metaphor for calmer times.

Fred's final report from the instructors at Laverton commented that, after flying heavy bombers, Fred was 'heavy' on the controls of lighter aircraft, such as the Oxford and the Anson. His flying of the Liberator was met with approval - smooth circuits and landings and good understanding of the cockpit. Imagine driving a road train, and then transferring to a Morris Minor!

The only incident recorded during this period was engine failure on a Liberator, with a return to base, and one perfunctory remark, "fire under the deck". It seems that a fire extinguisher operated by the second pilot would have taken care of that.

An unrecorded incident, now folk lore, occurred when Fred flew a small aircraft at low height, along Ruthven St, Toowoomba.

'Under the power lines!" some said.

Fred was grounded for a week.

November 1945

Fred's last flight as a R.A.A.F pilot was from Amberly, with an air test of his beloved Liberators. He did not know then how very hard it would be to stop doing what he so loved.

After the War

Fred returned home to Toowoomba, not sure what to do with himself. Although no longer on active duty, he was still chasing the paper trail of bureaucracy.

15th January 1946

Fred claimed 53 days of leave for overseas duties. As this would be pay in lieu, Fred's parents, concerned that he would simply spend his accrued cash funds, encouraged him to buy a small grocery store that was for sale a few blocks from the family home on the corner of Hume and Long Streets. The

Bright family had several generations of shop keepers, hence the inspiration. The corner store business became a family affair, as Bill and Clara moved themselves and their mostly adult children to the building, which included a residence behind the shop. Bill was still working for the railway, so the day to day running of the business was left mostly to the others. The shop would close at 7pm, meaning that one of the family would remain at the front counter, while the others had their 6pm dinner. Such dedication to Fred's well-being echoes the depth of love the family held for him.

After Fred was demobbed on 22nd June 1946, his restlessness was creeping up again. He started losing interest in the corner store leaving the family to run it. Youngest daughter Heather would help out after school. Eventually the business was sold, and the family moved again, to Simla, in Bridge St. The premises has survived, now being a hair salon at the front, and rental flats at the rear. The family was keen for life to 'return to normal' but Fred was constantly restless. Any talk about joining the Civilian Air Force, or piloting commercial aircraft brought a flood of tears from Clara.

"Just do something sensible!" she pleaded.

Fred said nothing, but continued planning.

10th February 1947

Almost three years after leaving 462 Squadron, Fred finally officially received his Commission. There was no fanfare, just a piece of paper.

"Still," he thought to himself, "it might help me get back in the air force."

During his service he had earned the following medals:

1939-1945 Star
Africa Star and Clasp
Italy Star
Defence Medal
War Medal 1939-1945
Australia Service Medal 1939-1945
Returned from Active Service Badge

Fred re-joined Gordon Motors, which had employed him as a bookkeeper. Although agricultural, teaching and industrial occupations were classified as 'reserved', commercial jobs were not included. In a relatively small country location post war, it was likely that patriotism for those who served overseas were given a first preference.

To friends and family, Fred was 'settling down' at last. Internally, he was unsettled, and thought about little else but flying.

1948

Fred met Neta, my mother, at a nurses' winter ball in Toowoomba. They became engaged six weeks later and married on October 30th. I was born July 1949.

Fred remained eternally restless and spent more time at the pubs than he should, much to the dismay of Neta, coping with a new baby.

30th June 1950

Fred had applied to join the Active Reserve of the R.A.A.F. He presented well at interview and was highly recommended as a pilot. His application was approved, and Fred was ranked as Flying Officer.

2nd February 1951

Fred was informed that he was not required 'at this time' by the Active Reserve, but his details were kept on file.

Bitterly disappointed, Fred joined Redman Motors, and his little family moved to Brisbane. Thus began the nomadic life we had during my childhood, as he changed his focus to his business interests, working his way up the corporate ladder.

12th December 1955

Fred received a letter, stating that he had been transferred from Active Reserve to General Reserve. This meant that his position was one of being on call, and non-operational. His disappointment at his dream slipping further and further away bit deeper.

He started drinking heavily to dull the pain, yet functioned in his daily life. At this time Fred, Mum and I were living in Lismore, New South Wales. Three years later he joined the Australian Guarantee Company and moved us back to Toowoomba, where my brother was born, May 1958.

During his career, Fred was transferred regularly to gain promotion. Most of those moves were by air, so Fred would take every chance to chat to the stewardesses, and gain permission to enter the cockpit. He would return to his seat with a huge grin on his face. Even just being inside a cockpit was enough to boost his morale.

Many, many model aeroplanes were made, of balsa, and plastic. At first it was for his own interest, but he taught my younger brother to build his own. Model aircraft kits became a popular birthday or Christmas present.

For the remainder of his life, Fred would either march on ANZAC Day, or go to the RSL to reminisce with ex-service men

No mention at home was ever made about Fred's burning desire to fly again. For the family, it was something he had done 'in the past'. For Fred, it was with him night and day.

1970

Fred, Neta, and my brother, Bill Junior had moved to Brisbane. I was nursing in Toowoomba. I heard years later that Fred had applied to join the active R.A.A.F, to help fight in the Vietnam War. Fred was then approaching fifty years of age, and his application was denied. Once again, the disappointment bit hard.

Fred had 'contacts' at the Archerfield Airfield in Brisbane, where he had first learned to fly Tiger Moths. A pilot friend would 'take him up' in a light Cessna aircraft and let Fred take the controls. Oh the bliss of being airborne again, with 'stick' and rudders in front of him.

1976

Fred passed away from cancer in October, with his service number engraved on his watch, as I had previously organised for him, a few weeks before. Right to the end, he treasured his air force status, and his life as a pilot.

Epilogue

Now that I have completed this book, I understand so much more about Dad than I did at the beginning. It is highly likely that he did suffer from the effects of post-traumatic stress, but my feeling now is that Dad, who fell in love with flying and aviation in his youth, never recovered from losing the opportunity to keep flying professionally, post-war. An occasional joy flight in a Cessna from Archerfield Airfield, his training ground, had to suffice. His parents' resistance to his staying on in active service would have added to that disappointment. Mum would not have been enthused with being an air force wife, but she would never have resisted, understanding Dad's unfulfilled passion.

Well Dad, I feel humbled by the opportunity to write about you. I hope I have done you service.

I can assure you that your daughter (the author) also has the flying bug. I do have some regret that I did not have the opportunity to learn how to fly, but my life on the ground has been rich and fulfilling. I satisfy my aviation passions by watching 'Twelve O'clock High' or reading one of my many books about flying. I instinctively look skywards when a Cessna, commercial or military aircraft flies over my house.

I prefer flying Virgin from Brisbane Domestic Airport, as I can watch Boeing 737s take-off and land from the departure gate. It still takes my breath away.

I am sorry that we never had the chance to talk about our mutual love. I think we would be pleasantly surprised to find that we had much more in common than we believed.

FLY HIGH, FRED!

1.

ABOUT THE AUTHOR

Penny de Jong has worked as a nurse specialising in community and palliative care, and as a volunteer for community groups. She has juggled these with her 'other' career as a wife, mother and grandmother.

Penny was a contributor to the book, *Menopause; Women tell their stories*, edited by Debra Vinecombe, Wakefield Press South Australia, 2008. ISBN 9781862547704 and also contributed the story Anna's Boys to the *ANZAC* Centennial Issue of Fellowship of Australian Writers (Qld), 2018.

Penny has a passion for travel, and inherited her father, Fred's, passion for airplanes and flying.

About the Editor:
Edwina Harvey is a writer/editor with over twenty years experience editing fiction and non-fiction including academic articles and theses. She gained her editing qualifications in 2012 and runs Edwina's Editing Services as a freelance editor. (edwinaseditingservices@gmail.com)
Edwina and Penny met when they were both travelling in Ireland in 2019.

Peggy Bright Books: proudly publishing square pegs for round holes. Find more of our titles at www.peggybright-books.com. Thank you for supporting Australian small press!

www.ingramcontent.com/pod-product-compliance
Lightning Source LLC
Chambersburg PA
CBHW070558010526
44118CB00012B/1373